J. M. HOOBAN

Research, Technological Change, and Economic Analysis

Research, Technological Change, and Economic Analysis

Edited by

Bela Gold
Case Western Reserve University

Lexington Books
D.C. Heath and Company
Lexington, Massachusetts
Toronto

Library of Congress Cataloging in Publication Data

Main entry under title:
Research, technological change, and economic analysis.

Includes bibliographical references and index.
1. Technological innovations–Addresses, essays, lectures. I. Gold, Bela.
HC79.T4R47 338'.06 76-50496
ISBN 0-669-01286-6

This book was prepared with the support of NSF Grant PRA 7518861 A01. However, any opinions, findings, conclusions, or recommendations herein are those of the authors and do not necessarily reflect the views of NSF.

Published simultaneously in Canada.

Printed in the United States of America.

International Standard Book Number: 0-669-01286-6

Library of Congress Catalog Card Number: 76-50496

Contents

Preface

The increasingly urgent problems of economic growth, inflation, international competition, and more effective resource development are forcing major reorientations in the analytical foci of economics. Some of the most important shifts are

1. from the national economy viewed as a coherent entity to a more penetrating parallel concern with its major industrial and regional components
2. from a virtually monopolizing concern with aggregate demand levels to a fuller consideration of the sources and composition of supply
3. from a primary concentration on the financial level of analysis (incomes, expenditures, and investments) to a complementary concern with the physical resource flows underlying them
4. from explicitly excluding technological innovations to including them not merely as an addition to the traditional inputs of labor, capital, and natural resources, but rather as an integral component of the interactions affecting the nature and price as well as the magnitudes of input requirements and output potentials.

This volume examines the last reorientation. But the ramifications of technological advances are so pervasive that a critical review of achievements and shortcomings in encompassing their economic effects cannot but engender spillover implications for dealing with the other needs as well.

That we know less about the economic effects of technological advances than would be desirable cannot occasion much surprise. Much the same can be said of all fields of knowledge. It is more important to seek answers to such questions as: How much of accepted beliefs and findings may be open to serious question? How valid are the concepts and tools being used to seek further understanding? Has the slow pace of progress in this field, despite its enormous importance for industrial and government policy, been caused only by the paucity of research efforts?

This critical review of the literature and its concluding proposals are the outcome of a series of studies made by the Research Program in Industrial Economics at Case Western Reserve University during the past ten years. These projects have yielded a substantial array of publications concerned with the economic effects of technological innovations in a variety of industries.[1] Moreover, the present volume represents only the first stage of a research project that provides for immediate field testing of the approach presented in the final chapter. The entire project has been financed by a grant from the National Science Foundation (PRA 7518861 A01). The opinions, findings, and conclusions presented, however, are those of the authors and do not necessarily reflect the views of the Foundation.

Although I designed and served as chief investigator of the project in addition to editing the reports, it should be emphasized that all contributions benefited from extensive joint discussions among the contributors.

Bela Gold

Cleveland, Ohio
October 1, 1976

Note

1. Relevant books include: B. Gold, *Explorations in Managerial Economics: Productivity Cost, Technology and Growth* (New York: Basic Books, 1971); D. Huettner, *Plant Size, Technological Change and Investment Requirements* (New York: Praeger, 1974); M.G. Boylan, *Economic Effects of Scale Increases in the Steel Industry* (New York: Praeger, 1975); B. Gold (ed.), *Technological Change: Economics, Management and Environment* (Oxford: Pergamon, 1975); and S. Eilon, B. Gold, and J. Soesan, *Productivity Analysis for Industry* (Oxford: Pergamon, 1976). A list of reprints and working papers is included in the *Progress Reports* of the Research Program in Industrial Economics.

Research, Technological Change, and Economic Analysis

1 Prevailing Beliefs and Analytical Needs

Bela Gold

The accelerating pace of technological advances is reshaping national resources requirements, the utilization of human resources, the location of industries, patterns of consumption, and environmental conditions. Such developments have engendered a growing volume of publications by economists eager to help guide efforts to cope with emerging problems. Unfortunately, however, little of it has proved pertinent to industrial executives who must choose among alternative innovational possibilities or to government officials groping for policies to encourage technological progress in directions likely to maximize the resulting welfare benefits. Why?

To help answer this question, the literature of the past twenty years has been analyzed from the standpoint of its applicability in assessing the actual effects of actual innovations rather than the theoretical effects of hypothetical innovations under artificial conditions. This entails analyzing what kinds of technological changes have been dealt with, what kinds of effects have been measured, and how resulting findings have been interpreted. Digging somewhat deeper, it becomes necessary to examine the restrictiveness of the assumptions made, the limitations of the methodologies employed, and the validity of the attempted generalizations.

Prevailing Beliefs

Although relatively little was known about the economics or the management of innovational undertakings even after World War II, the possibility of gaining substantial advantages through such efforts, combined with the threat of losing position through lagging behind, engendered an atmosphere increasingly favorable to action. And with it emerged a supportive rationale that was simple enough and apparently persuasive enough to gain widespread acceptance of the following crude components:

1. Technological innovations are inherently attractive, especially in terms of the economic rewards which are widely considered to be overriding in business organizations.[1]
2. Major technological innovations are generated primarily by increasingly complex and heavily financed research and development programs—with yields roughly proportioned to the resources applied.[2]

3. The major effect of such technological advances, aside from yielding new products, is to increase "productivity."
4. Rising productivity yields progressively lower unit costs.
5. Lower unit costs generate increasing profitability and growth.[3]
6. Increasing profitability and growth foster more research and development as well as faster diffusion of innovations—thus retriggering this entire cycle.

Such beliefs gained general encouragement from spectacular achievements in military and space undertakings and more direct reinforcement from some outstanding successes in industry. Further reassurance was provided by the broadly disseminated results of a small but expanding array of theoretical and empirical studies concerned with the economic effects of technological innovations. In this favorable atmosphere, industrial and government allocations for research and development (R&D) and for the diffusion of technological innovations expanded rapidly from the early 1950s until the late 1960s.

The growing momentum of acceptance led in turn to increasing criticism of those deemed to be insufficiently committed. Some firms and industries were chided for allocating only 1 or 2 percent of sales to R&D instead of 7 or 8 percent. Others were attacked for not adopting known process innovations fast enough. And such criticisms frequently were coupled with urgent recommendations for intensifying technological improvement efforts and introducing stronger governmental incentives and support for such programs. Accompanying arguments seemed to imply that the firms and industries which were their targets were either unaware of the attractive rewards available to them, or that they preferred avoiding change, even at the cost of foregoing substantial benefits.[4] There are, of course, bound to be some instances of one or another of these shortcomings at all times and perhaps in all economic sectors. But it would take considerable evidence to justify claims that managerial inadequacies of such gross proportions were pervasive throughout most of American industry, or even large sectors of it, for extended periods.

Accordingly, there is good reason to analyze in considerable detail all phases of the arguments and of the evidence underlying such critical judgments and the ameliorative policies proposed. Indeed, the desirability of doing so has been sharply accentuated in recent years, both by the reduced growth in R&D reported by government statistics (which also are presumed to reflect comparably lessened responsiveness to innovational improvements in general) and by the renewed pressures for reversing such trends (see, for example, "The Silent Crisis in R&D," 1976).

Analytical Needs

The initial and single most important focus of efforts to evaluate the economic effects of major technological innovations in industry must be the individual firm, for that is the context in which decisions are made to adopt or to reject prospective innovations on the basis of their expected results. In analyzing such effects, industrial managements are interested primarily in their bearing on the basic criteria of corporate performance and in identifying the means of engendering the most favorable and longest lasting results. Such objectives require approaches to the analysis of attendant changes in physical inputs and outputs that can be integrated into a larger analytical framework encompassing the determinants of company costs, revenues, investment needs, and profits.

The resulting combined framework would enable management to trace the effects of an actual or prospective innovation on inputs, processing methods, and product mix through the successive linkages—from procurement through production and distribution—that engender eventual changes in the performance of the plant or firm. And such a framework would also permit working backward from changes in aggregate performance to uncover the initiating innovations and subsequent interactions responsible for the outcome. Systematic applications of an analytical framework of this kind should provide management with sounder bases for appraising alternative means of improving production; for evaluating the probable effects of any given innovation on the level, quality, and mix of input requirements as well as on production costs and revenues; and for estimating the rate at which responsive adjustments in factor and product markets are likely to erode the early benefits of innovations.

But the effects of the technological innovations introduced by individual firms also tend to ramify outward in various directions, thereby introducing additional contexts within which such effects may have to be evaluated. Resulting changes in the quantity, quality, prices, and mix of inputs may engender significant repercussions in the industries and factor markets that supply them—including mining, fuel and power, and material processing industries (foreign and domestic), as well as capital goods producers and suppliers of investment funds. And resulting changes in the quantity, quality, mix, and prices of products may affect the industries producing competing products in addition to the industries or consumer groups that purchase these outputs. Moreover, the firm's innovation may initiate competitive pressures in factor and product markets on the other firms in its industry, as well as pressures for shifts in the industry's common strategies concerning relationships with trade unions, local and national government agencies, and overseas suppliers and competitors.

Governmental interests in evaluating the effects of technological innovations overlap most of the preceding contexts in addition to reaching beyond them. A variety of federal departments and agencies are concerned with the economic viability—including growth, technological progress, and cost competitiveness—of industries supplying inputs of materials, fuel, and power; and these plus others are also concerned with the development and effective utilization of natural resources and the maintenance of an attractive environment. Additional government agencies are concerned with employment levels, training needs, industrial relations, and the protection of health and safety in all industries. Still others are concerned with the maintenance of effective competition, our competitive position in foreign trade, the adaptation of capital markets to changing needs, the interactions between tax and regulatory policies and business policies, providing incentives to research, safeguarding military security, and even broader economic and social objectives. Thus, each of these government agencies may be concerned with evaluating the effects of past or prospective technological innovations—although some more comprehensively and more continuously than others.

Therefore, both industrial and governmental needs urge the development of an approach to the analysis of technological innovations that covers the sources and effects as well as the magnitude of changes in physical inputs and outputs; that traces attendant interactions among the physical, financial and welfare aspects of such flows over progressively broader sectors; and that also encompasses the changing pattern of such relationships through time. And economists face similarly broad arrays of analytical urgencies, including the need to develop more penetrating insights into the factors affecting the level of resources allocated to promoting technological advances; shifts in the targets of such efforts; variations in the rates of application of emerging advances; the widely varying responses to such advances in factor and product markets; and the speed with which the effects of innovations reach a peak and are then dissipated.

Even more specifically, few decisions in industry or in government can be focused on R&D in general, or on the diffusion of technological innovations at large, or merely on increasing or decreasing aggregate commitments for such purposes. Practical choices are much more likely to center around choosing some among alternative R&D projects; between adopting particular innovations sooner or later; and whether to expand, maintain, or contract undertakings involving different levels of costs, risks, and potential returns. Moreover, evaluations that purport to be relevant to policy guidance and future decision-making should encompass a wide coverage of results instead of concentrating primarily on successful outcomes. They should also cover attendant difficulties, burdens, and disadvantages. And such evaluations should also be made at levels of aggregation comparable to those at which decisions are made, because the average results of experience in six to ten industries may be inapplicable to any one of them.

It is against this framework of decision foci and evaluative criteria that the available literature on the economic effects of technological advances must be appraised if it is to provide sound and relevant guidance for the shaping of industrial and public policies. Such a critical review must first cover the basic concepts, the analytical tools, and the specific measures entering into the design of the models used. And the examination must then be extended to include the representativeness of the innovational activities covered by empirical studies, the adequacy of the collected data, the rigor of the analysis, and the persuasiveness of the resulting findings and interpretations.

Plan of Study

This volume presents the results of a critical review of the literature of the past twenty years relating to the economic effects of major technological innovations. Although such efforts seem to have had a variety of objectives, this review is concerned primarily with the applicability of the concepts, tools, and findings for appraising the effects of past and of prospective innovational undertakings in plants, firms, and industries under actual economic conditions. Because of this specialized focus, it would obviously be inappropriate to extend our conclusions to other analytical contexts without comparably detailed evaluations.

The three chapters of Part I appraise the strengths and limitations of the analytical approaches that have dominated research in this area. Robert Baird concentrates on applications of production functions and such variants as total factor productivity measurements. Asim Erdilek covers input-output analysis and some of its close variants. Gerhard Rosegger deals with the conceptual framework of approaches to analyzing the related problems of the diffusion of technological innovations.

In Part II, the emphasis shifts to examining the results of studies on actual innovations in industry. In Chapter 5, Myles Boylan concentrates on findings relating to the sources of technological innovations; in Chapter 6, he and his associates deal with the effects of such innovations as well as with factors found to affect their rates of diffusion. Both chapters are rooted in the past publications of this Research Program.

A second objective of this volume is to use the foregoing evaluations as the basis for proposing means of strengthening the available analytical framework to provide more effective contributions to meeting the policy development and decision-making needs that have been reviewed in the preceding discussion. Thus, Chapter 7 attempts to identify the sectors in which the available methodologies fulfill and fall short of such needs. And Chapter 8 offers an analytical framework that seems to be more fully in accord with operational needs of industry and government, as well as the theoretical and empirical requirements of more penetrating economic analyses.

Notes

1. For reflections of such business attitudes, see Holland (1958, Chapter 2); Larrabee et al. (1960, p. 170); and Keezer et al. (1960, pp. 355-69). Economists expressing similar views include Schmookler (1956, pp. 183-90); Carter and Williams (1957, p. 38); Nelson (1959, p. 101); Brozen (1960, pp. 204-17); Sutherland (1959, pp. 118-35), and Minasian (1962, pp. 93-141). But additional motivations are also considered by Carter and Williams (1958, p. 56); Nelson

(1959, p. 124); and Sutherland (1959, pp. 118-35). Those placing primary emphasis on other motivations include Keynes (1936, pp. 161-62); Andrews and Brunner (1952, pp. 355-56); and Hershey (1966, p. 262).

2. Proponents of this general view include Schumpeter (1950, p. 132); Maclaurin (1953, p. 104); Galbraith (1952, p. 9); and Machlup (1962, p. 153). Opposing views have been offered by Jewkes et al. (1959, pp. 28-32, 225-30, 237); Kuznets, (1962, pp. 47-48); Heald (1964, p. 1111); and Wiesner, (1963, p. 214). For an excellent general review, see Nelson (1959). For expectations and empirical findings that research outputs may be expected to rise on the average with increases in research inputs, see Silk (1960, p. 170); Ulin (1958, p. 27); Mansfield (1964, pp. 319-49, and 1965, pp. 310-322); and Comanor (1965, pp. 182-90), as well as Machlup (1962, pp. 152-53) and Griliches (1962, pp. 349-50). For judgments and findings questioning such relationships, see Jewkes et al. (1959, pp. 150-51, 225-26, 230, 237) and Nelson (1959, pp. 112-13, 118 and also June 1959, pp. 299,301). Other useful findings are provided by Kuznets (1962, pp. 47-48); Mueller, (1962, pp. 323-58); Marshall and Meckling (1962, p. 463); Klein (1962, pp. 478-80 and p. 508); Carter and Williams (1957, p. 19); Coales (1957, pp. 239-42); Rubel (1963, p. 35, 43); and von Braun (1963, p. 1080).

3. For evidence supporting expectations of increasing economic benefits from innovational efforts, including enhanced profitability and growth, see Mansfield (1964, pp. 319-40 and 1965, pp. 310-22), as well as Minasian (1962, pp. 100 et seq.) and Comanor (1965, pp. 182-90). For expressions of continuing doubt, see Griliches (1962, pp. 349-50); Hitch (1962, p. 193); Sanders (1962, p. 64); Mueller (1962, p. 344); Hershey (1966, p. 263); and Quinn and Cavanaugh (1964, p. 118). Also see the findings and comments of Enos (1962, p. 310); Markham (1962, p. 601); and Worley (1962, p. 234).

4. For a summary of published arguments and findings, see Gold (1971, pp. 211-22).

References

Andrews, P.W.S., and Brunner, Elizabeth, *Capital Development in Steel* (Oxford: Blackwell, 1952).

Brozen, Yale, "Trends in Industrial Research and Development," *Journal of Business* (July 1960).

Carter, C.F., and Williams, B.R., *Industry and Technical Progress* (London: Oxford University Press, 1957).

_______, *Investment in Innovation* (London: Oxford University Press, 1958).

Coales, J.F., "Financial Provision for Research and Development in Industry," *Journal of Industrial Economics* (July 1957).

Comanor, W.S., "Research and Technical Change in the Pharmaceutical Industry," *Review of Economics and Statistics* (May 1965).

Enos, J.L., "Invention and Innovation in the Petroleum Refining Industry," in *The Rate and Direction of Inventive Activity: Economic and Social Factors*, ed. R.R. Nelson (Princeton: Princeton University Press, 1962).

Galbraith, J.K., *American Capitalism* (Cambridge, Mass.: Houghton Mifflin, 1952).

Gold, B., *Explorations in Managerial Economics* (New York: Basic Books, 1971).

Griliches, Z., "Comment," in *The Rate and Direction of Inventive Activity: Economic and Social Factors*, ed. R.R. Nelson (Princeton: Princeton University Press, 1962).

Heald, Henry R., in U. S. Congress House, Select Committee on Government Research, *Summary of Hearings: Federal Research and Development Programs* (1964).

Hershey, R.L., "Finance and Productivity in Industrial Research and Development," *Research Management* (July 1966).

Hitch, C.J., "Comment," in *The Rate and Direction of Inventive Activity* (1962).

Holland, Maurice, *Management's Stake in Research* (New York: Harper & Bros., 1958), chap. 2.

Jewkes, J., Sawers, D., and Stillerman, R., *The Sources of Invention* (New York: St. Martin's, 1959).

Keezer et al., "The Outlook for Expenditures on Research and Development During the Next Decade," *American Economic Review, Papers and Proceedings* (May 1960).

Keynes, J.M., *The General Theory of Employment, Interest and Money* (London: Macmillan, 1936).

Klein, B.H., "Reply and Reminder," in *The Rate and Direction of Inventive Activity* (1962).

_______, "The Decision Making Problem in Development," in *The Rate and Direction of Inventive Activity* (1962).

Kuznets, S., "Inventive Activity: Problems of Definition and Measurement," in *The Rate and Direction of Inventive Activity* (1962).

Larrabie, W.M., et al., *Profit, Performance and Progress*, cited in *The Research Revolution*, by L. Silk (New York: McGraw-Hill, 1960).

Machlup, Fritz, "The Supply of Inventors and Inventions," in *The Rate and Direction of Inventive Activity* (1962).

Maclaurin, W.R., "The Sequence From Invention to Innovation and its Relation to Economic Growth," *Quarterly Journal of Economics* (February 1953).

Mansfield, E., "Industrial Research and Development Expenditures: Determinants, Prospects and Relation to Size of Firm and Inventive Output," *Journal of Political Economy* (August 1964).

_______, "Rates of Return from Industrial Research and Development," *American Economic Review* (May 1965).

Markham, J.W., "Inventive Activity: Government Controls and the Legal Environment," in *The Rate and Direction of Inventive Activity* (1962).

Marshall, A.W. and Meckling, H.W., "Predictability of the Costs, Time and Success of Development," in *The Rate and Direction of Inventive Activity* (1962).

Minasian, J.R., "The Economics of Research and Development," in *The Rate and Direction of Inventive Activity* (1962).

Mueller, W.F., "The Origins of the Basic Inventions Underlying Dupont's Major Product and Process Innovations, 1920-1950," in *The Rate and Direction of Inventive Activity* (1962).

Nelson, R.R., "Economics of Invention—A Survey of the Literature," *Journal of Business* (April 1959).

Nelson, R.R., (ed.) *The Rate and Direction of Inventive Activity: Economic and Social Factors* (Princeton: Princeton University Press, 1962).

Nelson, R.R., "The Simple Economics of Basic Scientific Research," *Journal of Political Economy* (June 1959).

Quinn, J.B., and Cavanaugh, R.M., "Fundamental Research Can Be Planned," *Harvard Business Review* (Jan-Feb. 1964).

Rubel, J.H., in U.S. Congress Senate, Committee on Small Business, Hearings (1963).

Sanders, B.S., "Some Difficulties in Measuring Inventive Activity" in *The Rate and Direction of Inventive Activity* (1962).

Schmookler, J., "The Level of Inventive Activity," *Review of Economic Studies* (May 1956).

Schumpeter, J., *Capitalism, Socialism and Democracy* (New York: Harper, 1950).

Silk, L., *The Research Revolution* (New York: McGraw-Hill, 1960).

Sutherland, A., "Diffusion of an Innovation in Cotton Spinning," *Journal of Industrial Economics* (March 1959).

Ulin, R.P., "Thinking Ahead: What Will Research Bring About?," *Harvard Business Review* (Jan-Feb. 1958).

U.S. Congress House, Select Committee on Government Research, *Summary of Hearings: Federal Research and Development Programs*, 88th Congress, First and Second Sessions, 1964.

U.S. Congress Senate, Subcommittee of the Select Committee on Small Business, Hearings: *The Role and Effects of Technology in the Nation's Economy*, 88th Congress, First Session, 1963, Parts 1, 2, and 3.

Von Braun, W., "Evaluation of End Results of Research," in Part 3 of the U.S. Congress Senate, Committee on Small Business, Hearings (1963).

Wiesner, J.B., in Part 2 of the U.S. Congress Senate, Committee on Small Business, Hearing (1963).

Worley, J.S., "The Changing Direction of Research and Development Among Firms," in *The Rate and Direction of Inventive Activity* (1962).

"The Silent Crisis in R&D," *Business Week,* March 8, 1976.

Part I
Analysis of Dominant Approaches

2 Production Functions, Productivity, and Technological Change

Robert N. Baird

Background and Purpose

Attempts to separate and measure the relationships among production, factor productivity, and technological change are most frequently centered on the concept of a unique, estimable production function. In its ideal form, this function shows the maximum amount of some physical output that could be produced with the technically efficient combinations of the relevant physical inputs. Although the production function has its conceptual foundations strictly in microeconomic units, a vast number of studies completed during the past two decades have incorporated the function at national and even worldwide levels of aggregation.

The popular historical roots of the production function are found in the writings of David Ricardo early in the nineteenth century, and by the time Marshall had written the eighth edition of his famous *Principles*, the microeconomic production function had been developed nearly to the level that is currently employed in most introductory textbooks. Marshall's mathematical appendix traced out the expected properties of the function, and his note on the law of diminishing returns explored the subtle problems that continue to vex current economic theorists. Among these problems are utilization rates, the problem of assigning a value to capital utilized, and the problem of aggregating both inputs and outputs to levels above simple processes.

In current microeconomic theory, the production function serves as the foundation of the theory of the firm, especially in the short run. The demand for variable inputs is found by differentiating the production function with respect to the inputs (for example, dQ/dL) and by multiplying the resulting marginal products by marginal revenue (either P or $P + (Q)dP/dQ$, depending upon whether competition is perfect or imperfect). The resulting variable, called marginal revenue product of the factor, forms the firm's demand schedule for that factor. Similarly, the competitive firm's supply of output schedule is found by multiplying the marginal factor input per unit of output (dL/dQ) by marginal input cost (either W or dW/dL). The resulting marginal cost schedule is the firm's supply schedule as long as price is greater than average variable cost.

Few economists believe that these concepts can be applied straightforwardly

Professor M.G. Boylan and Mr. Badiul Majumdar offered many valuable comments and criticisms on an earlier draft and suggested many sources that I would otherwise not have found.

to describe the operations of real firms in real markets. The importance of these concepts is found in their ability to describe the behavior of markets, not firms. The supply schedules for outputs and the demand schedules for inputs are theoretically summed over all the firms in the industry to yield industry supply schedules for outputs and industry demand schedules for the inputs. On the basis of knowledge about the shapes of these functions, and similar ones derived from the theory of consumer behavior, economists attempt to predict at the market level many of the effects of institutional changes, government legislation, and technological advances. For example, it directly follows even from such simplifications that increases in the minimum wage rate should promote unemployment, that rent controls should create housing shortages, and that the adoption of an innovation that raises the marginal product of labor should increase the wage rate paid to labor.

The point made here is that the production function is a useful abstraction that helps in understanding some of the important forces that guide the behavior of markets. Although this version of the theory of the firm is not realistic, Friedman and others have argued that realism is not required in a good theory:

> A hypothesis is important if it explains much by little, that is, if it abstracts the common and crucial elements from the mass of complex and detailed circumstances surrounding the phenomena to be explained, and permits valid predictions on the basis of them alone. To be important, therefore, a hypothesis must be descriptively false in its assumptions; it takes account of, and accounts for, none of the many other attendant circumstances, since its very success shows them to be irrelevant for the phenomena to be explained (Friedman, 1953, p. 71).

Though the theory of the firm may be good for describing markets, it may not be a good theory for describing real firms. Many of the attendant circumstances that are perhaps irrelevant to the behavior of the overall market may be crucial to some or many of the individual firms in the market. The real firm is not some entity that hires capital and labor and mixes them together magically and continuously to produce something called output. The real firm hires labor of varying qualities and skills and combines it with machines of different types and vintages to process raw materials into finished outputs of perhaps several varieties. The firm must determine the size of its production run in accordance with set-up costs, expected sales per period, costs of holding inventories, and shipping costs. The management of the firm must negotiate wages and work rules with the officials of the union. The accounting office of the firm must determine depreciation charges and devise strategies for minimizing taxes. The differences between the real firm and the theoretical firm are magnified even more if the real firm happens to operate in both international and domestic markets. Though understanding all these and other functions of the real firm may not be necessary to predict the price and output behavior of the total market, it probably is necessary to study these relationships to learn how an individual firm reacts to a change in its environment. (This is why courses in managerial economics differ from courses in microeconomics.[1] It probably also accounts for the lack of generality of theories of oligopoly, because many of these theories attempt to copy the behavior of real firms.)

The previous analysis is perhaps a bit belabored by the obvious, but the belaboring is intentional. Many of the difficulties in the use of production functions and productivity ratios are grounded in the attempt to extend an abstraction that is useful in analyzing some types of problems to areas of application where its use is simply inappropriate. We shall demonstrate the truth of this statement by exploring some of the recent work in these areas.

In the past few years, there have been several excellent review articles devoted to exploring the production function as a means of evaluating productivity and estimating technological change. Among these have been the exhaustive studies of Walters (1963), Nadiri (1970), and Kennedy and Thirlwall (1972). In addition, the books by Lave (1966), Brown (1967), and Mansfield (1968) explore selected topics in the area more intensively. A thorough bibliography in this field through 1971 is available from Bureau of Labor Statistics Bulletins 1226, 1514, and 1976. It is not the purpose of this report to provide another such review and thereby duplicate much of this prior work. The concern here is to provide a summary of the most important issues in the relevant areas.

The next section describes the characteristics most commonly attributed to production functions that are employed in empirical studies by economists. Some of the major uses of this and related approaches are then reviewed. Later sections contain the major criticisms traditionally offered of such studies, as well as the additional criticisms and comments of this research staff.

Characteristics of the Microeconomic Production Function

The production function is commonly used in much empirical work related to the behavior of different levels of the economy.[2] The usual form of this function is:

$$Q = f(K,L) \tag{2.1}$$

where: Q = quantity of output in physical units, or an index of physical units
K = the capital stock, or services derived from it
L = labor services employed over period of consideration

It is customary to assume that the marginal products of each input are always positive, but that at some point the derivatives of the marginal product with respect to the inputs are negative because of the law of diminishing returns.

Moreover, another common assumption is that an increase in either of the two inputs will increase the marginal product of the other input. (Of course, this latter effect could be either active or passive. That is, the introduction of a machine in a production process could force the hiring of a more highly skilled operator than was previously employed. On the other hand, the introduction of the machine could allow labor of the same quality as before to produce more output.)

The concept of scale in the production function is considered by the mathematical property of homogeneity. Specifically, it is assumed that the function is homogeneous of some degree w, which implies that:

$$f(\nu K, \nu L) = \nu^w f(K, L) \tag{2.2}$$

If w is 1.0, the function exhibits constant returns to scale, and a doubling of both inputs will cause a doubling of output. Conversely, there are either economies or diseconomies of scale, depending upon whether w is respectively greater than or less than unity.

Another important characteristic of the production function is the degree of substitution which it exhibits. Specifically, the elasticity of substitution is the percentage change in the ratio of the factor inputs divided by the percentage change in the marginal rate of substitution of those two factors. Of course, because the marginal rate of substitution will be equal to the ratio of the input prices *at the point of most efficient use*, the elasticity of substitution can also be expressed with the percentage change in the ratio of the factor prices. If the elasticity of substitution is zero, then the isoquants of the production function appear as nested right angles when graphed in the K, L plane. Such functions are said to be fixed coefficient production functions because they imply that a certain amount of K (or L) is required to produce a certain amount of Q, no matter how much L (or K) is available, and inputs may not be substituted for each other. Input-output production functions and many different types of process functions are of this type. On the other hand, an elasticity of substitution of infinity implies that the isoquants are negatively sloped straight lines. In this case, inputs may always be freely substituted for each other in some fixed ratio with never a diminution of output. Obviously, we would expect empirical functions to fall somewhere between these two extremes.

An additional feature of the production function is the implication that it carries for factor shares. If factors are purchased in competitive markets and paid the value of their marginal products, then the production function is a determinant of how large a share each factor will receive. At the microeconomic level, this implication may not be important, but it increases in importance as the level of aggregation increases.

Several implicit assumptions underlying most production functions are related to the materials being processed. As shown above, the production function indicates that output is a function only of capital and labor, and materials are customarily excluded. For this to be legitimate, it must be true that such materials are available in perfectly elastic supply and that they are linked with output by fixed technical coefficients of production. In other words, materials cannot be substituted for capital and labor or vice versa. Though a later section is reserved for criticisms of traditional methods, this research staff believes that such treatment accorded to materials constitutes a major failure of these methods. We will demonstrate by showing examples where materials and capital are substitutable for each other.

The final important characteristic of the production function concerns the way in which technology is treated. At present, there are two basic approaches to handling this variable.

In the first approach, technological progress is said to be disembodied from the factor inputs. Somehow, according to this view, an increase in technology enables the same type of capital and labor to produce more output, even if total inputs do not increase. In other words, the production isoquants are shifted upwards by technology. Such production functions are written in general terms as:

$$Q = f(K, L, A(t)) \tag{2.3}$$

where: $A(t)$ expresses the growth of technology as a function of time

Three standard forms of disembodied technological progress are customarily studied.[3] In the first, the rate of technological progress is thought not to affect the marginal rate of substitution between the two inputs, and the production function is written as:

$$Q = f(K, L)A(t) \tag{2.4}$$

Such a function defines *Hicks-neutral* technological change.

A second view suggests that technological change makes capital more efficient even if that change is not embodied in new types of capital. Thus, the effective capital stock can be defined as:

$$K^* = A(t)K \tag{2.5}$$

and

$$Q = f(K^*, L) \tag{2.6}$$

This type of technological progress is called *Solow-neutral.*

Finally, it may be argued that technological progress augments the effectiveness of the labor stock in a manner analogous to the capital-augmenting Solow-neutral change. In the labor augmenting case, technological change is said to be *Harrod-neutral*, and the production function is

$$Q = f(K, L^*) \tag{2.7}$$

The second major approach to the inclusion of technology considers that technological advance is "embodied" in the productive factors, especially capital. As Solow (1960, p. 91) has stated:

> The striking assumption [of older models, including his own] is that old and new capital equipment participate equally in technical change. This conflicts with the casual observation that many if not most innovations need to be embodied in new kinds of capital equipment before they can be made effective.

If the rate at which technical progress is embodied in capital is constant, the capital stock at any time t is effectively a weighted sum of all past capital investments and thus could be expressed by the following integral:

$$K_t = \int_o^t [e^{rv - d(t-v)}]\, I_v d_v \tag{2.8}$$

where: d is the rate of obsolescence
r is the rate of technical change
I_v is investment made in period v

Of course, a similar equation could be developed to derive an effective labor stock to account for technological change that was partially embodied in labor.

If the production function is to be used to measure productivity or to evaluate the effects of technological change, the form of the function must be exactly specified before the estimation of the parameters can commence. The most common type of function used for such a purpose is the Cobb-Douglas function:

$$Q = AK^a L^b \tag{2.9}$$

where: A is an index of efficiency (cross-sectional studies) or technology (time-series studies)
a and b are constants

This function is linear in the logarithms of the variables, so the coefficients a and b respectively represent the elasticities of output with respect to capital and labor. The sum of a and b is the degree of homogeneity of the equation, and the elasticity of substitution is unity.

A second common form of specification is the constant elasticity of substitution (CES) function:

$$Q = A\,[fK^{-r} + (1 - f)L^{-r}]^{-1/r} \tag{2.10}$$

where: f is the distribution parameter and is a determinant of the relative factor shares in total output.
r is the substitution parameter and is a determinant of the elasticity of substitution which is constant, but not necessarily equal to unity

Because the elasticity of substitution can take on different values, the CES function is a strong theoretical improvement over the Cobb-Douglas function, although it is obviously more difficult to manipulate mathematically.

Both the previous functions are homogeneous with respect to the factor inputs, which presents a problem in the application of these functions to the level of the competitive firm. In competitive product markets, a production function that is homogeneous of the first degree implies a profit function that is also homogeneous of the first degree. Thus, if a profit of some amount exists for one particular combination of factors, then any amount of profit is theoretically possible simply by changing the scale of operations by an appropriate amount. Accordingly there is no solution to the profit-maximizing problem, and equilibrium on the part of the firm is indeterminate without additional information. These arguments are pursued in both Henderson and Quandt (1965) and Chiang (1974).

More recently, a variable elasticity of substitution (VES) function has been developed, but at present many of its properties have not been fully explored, and there is probably no advantage in using this function over the two previously mentioned.

Applications of Production Functions

The Process, the Plant, and the Firm

Prior to the midfifties, most of the work in applying production functions to microeconomic units was at the level of the simple process. A good example of the process type of function is given by Heady and Dillon (cited in Walters, 1963), who showed the output of corn per acre as a nonlinear function of pounds of two different types of current inputs, nitrogen and phosphate fertilizers. Thus, for a fixed output, they could determine the optimal amounts of the two fertilizers, given their prices. Unfortunately, this purely technical relationship could not be used to show how much corn should be produced, how large the farm should be, what amounts of capital and labor were required, or how farm activities should be coordinated, either spatially or over the long run. Finally, technology was omitted from the function altogether.

A similar approach was given in Chenery's prior attempt (1949) to develop a production function for pipeline transmission of gas from engineering data. By relying on engineering information alone, he was able to show possible trade-offs between two types of *capital* inputs (pipe size and pump size). When he combined the results with cost data, he was able to estimate long-run average cost curves that had negative slopes throughout. Again, however, by relying on a single process, he avoided such questions as how much gas should be pumped. The study is essentially an *ex ante* view of production and cost schedules, and it may give an indication of what type of transmission network a firm would choose to build with a given technology and a set of expected future prices.

A.R. Ferguson (1951) attempted a similar study by developing an engineering or technical production function for air transport. His end result was a function relating ton-miles (output) to fuel, labor, and a vector of capital services.

Process functions and their implied cost schedules have also been used to explore the existence of economies or diseconomies of scale related to plant size. Typical of these are Johnston (1952), Nordin (1947), and Lomax (1952). Each of these is reviewed intensively and added to by Galatin (1968). Walters (1963, p. 39) has pointed out that cost functions are more convenient to employ than production functions because accounting data are most frequently presented in dollar terms. Indeed, he argued that it was the only possible method to use in some cases. Much of the theory and applied work in this area is given in Johnston (1960). A very recent work (1975) is Boylan's study of scale increases of blast furnaces, an imaginative study that also explicitly takes account of changes in the quality of raw materials being processed.

There are very few studies of production or cost functions for individual firms because of the great difficulty of compiling an adequate time series of consistent data. Sahota (1966), attempted to use *Moody's Industrial Manual* to develop enough data on nine chemical firms so that he could evaluate the relative efficiencies of each and estimate their rates of technological advance over time. By running twenty-two regression equations for three different time periods, he concluded that all nine of the firms were of similar efficiency, and that two-thirds of the increase in output over the 1936-60 period was caused by "technological change." Sahota's methods are clever, but not very convincing because of the very extensive manipulation of fairly crude data.

Sherrard's study (1967) of a single lumber company relied on physical output and input measures employed by the firm, but his conclusions (such as output per man hour is positively associated with log size) were usually obvious. Moreover, to consolidate many different outputs into one measure he relied heavily on "equivalent board feet," a measure of dubious economic meaning. Sherrard attempted no explicit measure of technological change.

The Industry

The most commonly studied microeconomic unit is the industry, frequently classified by the two-digit SIC code or its foreign equivalent. The obvious reason for concentration on the industry is the availability of official government data periodically. The principal mode for investigating the behavior of production in the industry has been the Cobb-Douglas function, first estimated in 1927 (Cobb and Douglas, 1928), and popularized during the thirties by researchers such as Douglas and Bronfenbrenner (1939). Until high-speed computers were available, most of the early studies simply measured the coefficients of the two input variables to determine relative factor shares or to see if an industry was characterized by economies or diseconomies of scale. Walters (1963), has reviewed the problems in and the findings of about thirty such studies completed before 1963.

The development of modern computers and the publication of Solow's famous articles in 1957 and 1960 significantly altered the type of study to which industry data were subjected. Once Solow had shown that it was possible, he thought, to piece together estimates of technological change at the national level, the door was thrown open for the widespread application of similar techniques to the rich vein of industry data.

Over the past twenty years, the rate of technological advance in two-digit SIC classifications for both embodied and disembodied technological progress has been estimated for Cobb-Douglas, CES, and VES types of functions. We cannot enumerate such studies, which has been done adequately in the reports cited in the introduction. The foci of these studies have concentrated generally on (a) the determination of relative factor shares (e.g., Moroney, 1967); (b) the magnitude of the elasticity of substitution (e.g., Lovell, 1969); (c) the rate of technological change (C.E. Ferguson, 1969); and (d) the existence of economies of scale (Murti and Sastry, 1957). These studies are yet to produce a consensus on any of these foci, and many of the results are openly inconsistent.

The National Economy

Although a technological innovation is obviously introduced at the micro level, much of the impetus for research in the area has come from studies at the national level of aggregation. The first attempt at estimating the rate of technological change was Tinbergen's Cobb-Douglas function (1942) with an exponential trend applied to data for Germany, Great Britain, France, and the United States. The most widely acknowledged benchmark, however, is Solow's 1957 study, in which he estimated an efficiency parameter of a Cobb-Douglas function for every year from 1909 to 1949, using aggregate data for the United States. He then calculated the rate of growth of that parameter and called it the rate of growth of technology. Solow's method was essentially to isolate increases in deflated national income that resulted from changes in labor and capital and to designate any residual as due to technological progress. Although that technique is simple and the data fairly crude, the paper sparked a wave of imitation and extension.

The next major development in the utilization of aggregate production functions was also Solow's, when he proposed a capital-embodied measure of technological change in 1960. By combining the concept with a Cobb-Douglas function, he concluded that technological progress had advanced at about 2.5 percent per year for the years 1919-1958 (1960).

Following Solow's 1960 paper came the next major development—the formulation of the CES function by Arrow, Chenery, Minhas and Solow (1961). The parameters of this function are most commonly estimated by a two-stage multiple regression process. The function has been employed at the aggregate level by Beckman and Sato (1969), C.E. Ferguson (1965), and many others. The findings in these studies are not always mutually consistent.

Related Applications of Productivity Ratios

The studies cited above are based on explicitly stated production functions, but another group of studies attempts to evaluate technology and efficiency on the basis of productivity ratios where the productive process is at best implicit in the formula for the ratio. Some of the major ratios in this approach are reviewed and criticized below.

Total Factor Productivity

The name "total factor productivity" has been given, perhaps too carelessly, to a family of ratios that purport to measure changes in efficiency or technology. The two most prominent members of the family are the arithmetic index most commonly associated with the work of Abramovitz (1956) and Kendrick (1961). The other is the geometric index from Solow (1957).

The Arithmetic Index. This index assumes that the ratio of successive outputs may be found by multiplying the sum of the weighted ratios of successive years' inputs by an appropriate efficiency variable. The weights selected are prices of the inputs for some base year. As a formula:[4]

$$Q/_2Q_1 = A^*[w_0(L_2/L_1) + i_0(K_2/K_1)] \qquad (2.11)$$

where: Q_i = output in the ith year, or value of output
A^* = efficiency variable
L_i = labor input in ith year
K_i = capital input in ith year
w_0, i_0 = input prices in base year

Because the efficiency variable is the one that is unobservable, the formula is normally written with that variable on the left side:

$$A^* = (Q_2/Q_1) / [w_0(L_2/L_1) + i_0(K_2/K_1)] \quad (2.12)$$

The rate of advance of technology is thus $A^* - 1$. Unfortunately, the formula is not suited to measure this.

Among the many problems of the arithmetic index is the fact that suitable measures for many of the variables are simply not available. More will be said of this in the next section. In addition, it is not sensible to continue to use base year weights for the input ratios indefinitely. Over time, shifts of the capital-labor ratio will change the marginal productivities of the two factors, and it would seem that the formula should take cognizance of this fact. Kendrick (1961) has attempted to change the weights according to a smoothing function, and Maddala (1965) has used weights from annual regression coefficients. Both Domar (1962) and Lave (1966) have examined the issue, and they have concluded that satisfactory resolution of this problem is impossible. The arithmetic index "correctly" measures technological change only if the capital-labor ratio and the ratio of input prices remain constant.

The Geometric Index. Solow's geometric index is based on the implicit assumption of a linearly homogeneous production function with neutral disembodied technological change. The resulting formula for the rate of technological change is thus:

$$dA/A = dq/q - w_k(dk/k) \quad (2.13)$$

where: q = output per unit of labor
k = capital-labor ratio
w_k = elasticity of output with respect to capital

Nadiri (1970) has shown that this index and the Kendrick index give similar answers for small changes in the underlying variables. Extensive criticism of this approach is reserved for following sections.

Farrell's Index

In 1957, Farrell offered an index which he thought could be employed to measure the relative productive efficiency of the firm. Over time, changes in the index would measure the improvement or decline in the firm's efficiency or, if the firm were as efficient as possible in all periods, the rate of technological advance. The explanation of this index for the two-input case is aided by reference to Figure 2-1.

In Figure 2-1, point *A* represents the actual quantities of capital and labor employed to produce 10 units of output, while the isoquant shows the combinations of those inputs that could produce the same output if maximum efficiency were employed. The ray *OA* gives the combinations of *K* and *L* that have the same capital-labor ratio as point *A*. Farrell states that the "technical efficiency" of the firm is measured by the fraction *OB*/*OA*. Under the assumption of constant returns, this fraction gives actual output as a percentage of the potential ouput if maximum efficiency were attained. If the line *PP′* indicates the relative price ratio of capital to labor, the firm is also employing the incorrect capital-labor ratio, so Farrell defines the fraction *OD*/*OB* to be the "price efficiency" of the firm, because the costs of operating at point *C* are only that fraction of the total costs at point *B*. Finally, if the firm kept its technical efficiency constant while moving to the same capital ratio as at point *C*, its costs would be reduced by a factor *OD*/*OA*, which he defines as "overall efficiency."

In applying and extending his framework to an actual industry (U.S. agriculture), Farrell concentrated on measuring only the first of these because he thought that technical efficiency was the most meaningful. "Price efficiency is a measure that is both unstable and dubious of interpretation; its virtue lies in leaving technical efficiency free of these faults, rather than in any intrinsic

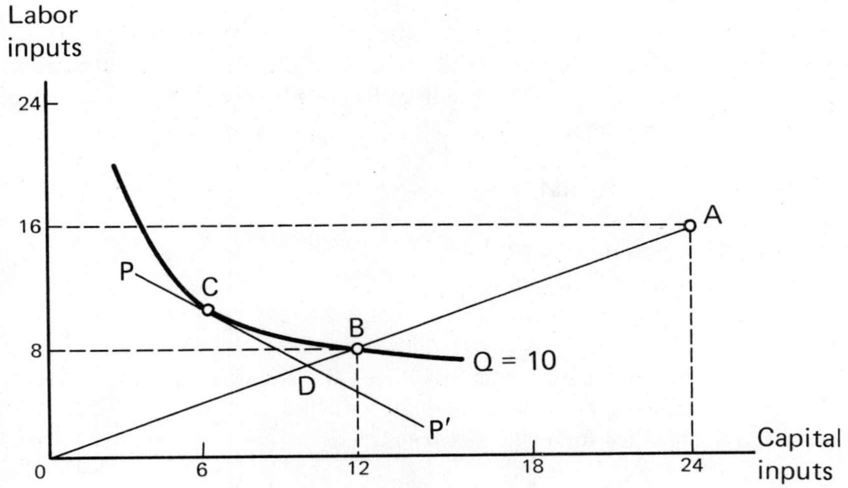

Figure 2-1. Farrell's Graph of Technical and Price Efficiency

usefulness" (Farrell, 1957, p. 261). To estimate the isoquants, Farrell bypassed the standard approach of statistically estimating the parameters of any prespecified function; instead, he pieced together the isoquants by examining linear combinations of different points on the production surface and selecting the lower bounds. As an illustration, he calculated technical efficiency indices for agricultural production in each of forty-eight states for 1952. In the illustration, he considered production processes that included up to four different inputs, and his results were closely correlated to the 1939 figures cited by M.G. Kendall (1957, p. 86) except for six states.

Farrell's method avoids some of the problems of relying on specific production functions, but it is nonetheless subject to some serious limitations. First, under the case of constant returns to scale, the technical efficiency index is identical to the Kendrick-Abramovitz arithmetic index mentioned earlier. As a consequence, this facet of Farrell's index shares the weaknesses of that index. Second, although no specific function is required for the estimation of the Farrell index, the data requirements are as great as for the approaches cited earlier. Thus, Farrell's measure still faces the problems of heterogeneity among inputs, defining and measuring capital, the aggregation problem, and so on. Third, the measure is sensitive to the number and type of inputs included. For example, his measure of agricultural technical efficiency in North Dakota was 55 percent when land and labor were the only inputs considered, but it rose to 100 percent when materials (in dollar value) were added. To the extent that inefficiency reflects omitted variables, Farrell's index is not appreciably better than Solow's residual. Fourth, because the number of possible linear combinations of observed input-output points increases dramatically as the number of inputs and firms under consideration is expanded, it was necessary in 1957 to restrict this number greatly. Advances in computer technology have mitigated but not eliminated this problem. Fifth, Eilon and Teague (1973) have shown that Sahgal's attempt (1972) to apply Farrell's measure to eleven single-product firms ended with results that were not promising, even in the two-input case. Finally, Farrell himself admitted that variations in the quality of inputs would reveal themselves as changes in efficiency. In view of all these problems, it is apparent that the Farrell index is not likely to be suitable for estimating technological change and its effects.

Productivity Accounting

In 1955, Hiram Davis suggested a method for determining changes in productivity that accounted for both different inputs and for changes in the composition of output and inputs over time. Basically, his approach is to evaluate outputs in terms of base year prices and to divide that total by the value of inputs in base year prices. This method of weighting is euphemistically referred to as "repricing," but the name does not disguise the fact that it is essentially another version of the arithmetic index applied to the level of the firm.

Several other indices could be reviewed here, but the end result is invariably the same: the weighting system involves the selection of arbitrary weights that cannot be permitted to remain constant for long periods of time. However, if the weights are changed every year, the formulas are reduced to simple accounting identities. In a following section, we will explain what we believe to be the essential errors common to all these productivity approaches and offer what we believe to be a sensible remedial policy.

Value Added Productivity

Although output per man-hour remains the most widely cited measure of changes in productivity, and of comparative productivity levels among similar industries in different countries, there has been growing awareness of its obvious shortcomings in general business useage. This has led to increasing resort in *Business Week, The Economist* (London), and other business journals (see, for example, Ball, 1968) to another ratio that is also simple but seems to be more sophisticated: value added per man-hour.

Unfortunately, the improvement offered by this ratio is small relative to its continued misrepresentation of the relative "efficiency" or "productiveness" of either labor alone or of the total operations covered. Specifically, this measure comes closer to representing the output attributable to the productive contributions of the given firm (or industry) because it excludes the contributions to the final products made by purchased materials and supplies. These latter frequently account for more than 50 percent of the total value of products that is reflected by measures of total output.

But this apparent gain is also accompanied by certain new as well as continued shortcomings. For example, value added per man-hour comparisons are heavily influenced by differences in the quality and level of fabrication of purchased materials, differences in input factor proportions because of differences in technology and in the specific types of products made, and by the effect of product differences on their prices. And, of course, this measure continues to ignore the fundamental fact that value added, like output, results from the joint contributions of many inputs rather than of labor alone. Hence, increases in value added per man-hour often result from reducing labor's contributions to output, through shifting some of its functions to machinery or other inputs, rather than from increases in the efficiency or productive contributions of the average man-hour.

Traditional Criticism of the Production Function as a Means of Evaluating Technological Change

Relevance of the Concept of the Production Function

As mentioned earlier, the production function is basically an abstraction that permits economists to explain the behavior of individual markets, and several people have suggested that it cannot really be used to describe real economic units.

To reduce the complexity of using production functions to analyze real production processes, most economists omit the inputs of materials from the function. Brunner (1974) has argued that this can only be done if materials are homogeneous and exist in perfectly elastic supply, an assumption that she rejects at most levels of aggregation.[5]

Second, the theoretical production function refers to maximum output that is attainable with given input combinations, yet empirical studies are necessarily devoted to the analysis of actual practice, not "best" practice. Klotz (1966) argues that there is at least a seven-year lag before best practice spreads through industry, and the entire literature on diffusion reinforces his contention.[6]

Winston (1974) has contended that capital is purchased with the idea that it will intentionally be idle a large portion of the time as a result of rhythmic variations in the prices of the variable factors. In addition, some idleness will be caused by a deficiency of demand for the products that are produced by the capital. He believes that only the latter type of idleness can be properly adjusted to measure the flow of services from the capital stock, and he is therefore pessimistic about the use of production functions that include capital:

> Since the utilization of capital is an economic variable, an important if obvious implication is that there can be no unique relationship between a capital *stock* and output or employment. Production functions defined on capital stocks are far more suspect than has generally been recognized.

Robinson (1953-54) and Kaldor (1962) have also argued against the existence of a production function based on a capital stock. Their arguments are essentially that above simple production processes, capital can only be measured by a value-related concept. However, the value of capital is the sum of its discounted stream of future net revenues, a sum that will vary as interest rates and price expectations are altered. Accordingly, there is no unique capital stock and no production function![7] Samuelson (1962) has contrived a production function that permits "reswitching" at the microeconomic level but still has Cobb-Douglas properties when aggregated to higher levels. Samuelson's model, however, makes some highly questionable assumptions:

1. Input coefficients are fixed at the micro level.
2. There is only one consumption good.
3. Constant returns to scale.
4. Perfect competition.
5. Only the *n*th capital good is used to produce the *n*th capital good.
6. Depreciation of a capital good is independent of its age.

Such restrictions clearly do not fall in Friedman's category of eliminating "irrelevant circumstances," and may even go so far as to create relevant (and perhaps necessary) ones. Lave (1966) contends that Samuelson's model is not likely to have the generality of application sufficient to overcome Robinson's objections.

Finally, Ames (1961) has warned that the production functions employed by economists may not be useful to either businessmen or engineers. He contends that usefulness requires that the production function show (a) what relationships are truly constant and (b) where better results can be obtained. Because estimated functions at best show a historical picture that has been forced into the academician's framework, they meet neither criteria, according to Ames.

Problems of Defining and Measuring Variables

Virtually every review paper on productivity has mentioned the difficulty of measuring the output of a production function when more than one final product is concerned. The basic problem is that different products can be added only on a dollar basis, and changes in the dollar figure from year to year, or from firm to firm, may reflect different market structures, (see Walters, 1963, p. 42), changes in quality of the output, or changes in product mix. Griliches (1964) has proposed a set of hedonic price indexes to attempt to adjust for quality changes, but the concept of quality has too many dimensions for such an approach to be fruitful for most products. There simply is no way of expressing, say, a 1976 Ford in terms of "Model-T equivalents."

Because the labor input is homogeneous neither within firms, across industries, nor over years, the measurement of labor by simple man-hours or related measures is not satisfactory, as pointed out by Nadiri (1970), Brunner (1974), and others. Attempts at making quality adjustments have been made by Jorgenson and Griliches (1962) and by Denison (1962). But, as Nadiri (1970, p. 1169) points out, all the adjustments embody large and arbitrary elements of choice in them.

The measurement of capital presents special problems because, as stated earlier, its value is itself an economic variable. Most studies that include a capital component in the production function, therefore, rely upon the cost of capital as its measure. According to Walters (1963, pp. 23-24):

The normal procedure is to measure capital value and then deflate by an index of prices. The ideal price would reflect the changing value of capital services. Unfortunately such a series is not available, and, instead, investigators have deflated by an index reflecting changing *costs* of labor and other inputs used in making capital. As Anderson has shown, increased productivity of labor, etc., in the industries producing capital goods is not incorporated in the price index. Thus, the increase in the quantity of capital is considerably understated.

Walters and others have argued that the proper measure for capital should correspond to the services of capital, but that such a measure is virtually impossible. Winston, as stated above, has shown that simple adjustments in utilization rates do not capture planned idleness, and Walters contends that subtracting depreciation merely corrects for the age of capital instead of its economic efficiency. In evaluating the treatment of capital, Walters concludes that "critics of empirical production functions are justified in expressing some skepticism of results which depend upon such unsatisfactory foundations" (1963, p. 24).[8]

The definition of technology has also been the source of much controversy in the literature. Because technological change cannot be measured by any conventional yardstick, its effect is commonly deduced by first accounting for everything else in the production function. The effect of technology will therefore be included in any discrepancy between what is accounted for by the known inputs and the actual output. Because of this, the rate of technological advance is often referred to as the "residual."

Solow notes that he is "using the phrase 'technical change,' as a shorthand expression for *any kind of shift* in the production function. Thus, slowdowns, speedups, improvements in the education of the labor force, and all sorts of things will appear as 'technical change.' "

Domar thought that a more appropriate name was the "residual"; that is, that part of increased output per man which is left over after increases in capital per man are accounted for (Lave, 1966, p. 5).

Nadiri's review article (1970, p. 1141) points out the consequences of treating technical change as a residual by saying that "any misspecification of the aggregate production function—errors in measuring variables, errors due to omission of the relevant inputs—will spill over to the measure of total factor productivity."

Limitations of Estimation Techniques

The estimation of the coefficients of production functions involves a set of tools whose growth has far outpaced the ability of the profession to use them wisely. Since the first crude multiple regression equations were estimated, new techniques for handling autocorrelation, heteroscesdasticity, simultaneity, and non-linear functions have all been developed and refined. Unfortunately, these refinements have not solved all the problems. Nadiri, for example, mentions that time series estimates of parameters will be biased because of simultaneity between inputs and prices, by misspecification of adjustment lags, and by cyclical conditions (1970, p. 1152). He also points out (p. 1155) that the new techniques used to estimate the CES parameters involves rigid assumptions:

> that marginal productivity conditions hold, which in turn requires that the returns to scale parameter is unity. Furthermore, it assumes that the responses of K and L to changes in relative prices . . . are identical, that the marginal productivity conditions hold instantaneously, and that the current relative prices are exogenous and good proxies for their expected values.

Lave has argued that the two-step process for estimating CES functions (and presumably for VES functions) requires more significant digits than are available in our existing data. He asserts: "Isn't it likely that if estimation requires the inversion of a large matrix, or complex adjustment of data by estimates from previous calculations, that the result will have the character of random numbers?" (1966, p. 141).

In summary, though the techniques are available, it is apparent that available data and the structure of our economic markets force us to look upon the estimated coefficients with great reservations concerning what they really measure.

Effects of Aggregation

If the microeconomic production function exists, if data can be properly defined and adequately measured, and if estimation techniques can furnish unbiased and consistent estimates of parameters, it is still not apparent that production functions at higher levels of aggregation exist. Can different outputs, inputs, and functional forms be combined in a manner so as to produce a single aggregated production function that has the properties previously described? Rather than review the mathematical and statistical arguments, let us consider the findings of some economists who have examined the problem intensively.

> After surveying the problem of aggregation, one may easily doubt whether there is much point in employing such a concept as an aggregate production function. The variety of competitive and technological conditions we find in modern economies suggests that we cannot approximate the basic requirements of sensible aggregation except, perhaps, over firms in the same industry or for narrow sections of the economy.[9] (Walters, 1963, p. 11.)

Aggregate measures of national product, productive wealth, and an employed labor force can be subjected to endless streams of criticism. Other series such as the return to capital, the share of capital in output, and price indices offer greater problems. A succinct comment on these problems is offered by Arthur Grant (1963). Although most economists agree that the share of labor in output rose from .43 to .63 between about 1890 and 1955, a careful reexamination of the Department of Commerce series led him to the conclusion of no change. If reasonable research can support the claim that this basic series could be mismeasured across such a large change, what confidence can be placed in the aggregate functions? (Lave, 1966, p. 141.)

The conclusion to be drawn from this brief discussion is that aggregation is a serious problem affecting the magnitude, the stability, and the dynamic changes of total factor productivity. We need to be cautious in interpreting the results that depend on the existence and specification of the aggregate production function. . . . That the use of the aggregate production gives reasonably good estimates of factor productivity is due mainly to the narrow range of movement of aggregate data rather than the solid foundation of the function. In fact, the aggregate production function does not have a conceptual reality of its own; *it emerges as a consequence of the growth processes at various microeconomic levels and is not a causal determinant of the growth path of the economy.* (Nadiri, 1970, p. 1146.)

In this day of rationally designed econometric studies and super input-output tables, it takes something more than the usually "willing suspension of disbelief" to talk seriously of the aggregate production function. But the aggregate production function is only a little less legitimate a concept than, say, the aggregate consumption function, and for some kinds of long run macromodels, it is almost as indispensable as the latter is for the short run. As long as we insist on practicing macro-economics, we shall need aggregate relationships. . . . Before going on, let me be explicit that I would not try to justify what follows by calling on fancy theorems on aggregation and index numbers. Either this kind of aggregate economics appeals or it doesn't. Personally, I belong to both schools. If it does, I think one can draw some crude but useful conclusions from the results. (Solow, 1957, p. 349.)

The theorems on index numbers and aggregation that Solow refuses to evoke pretty well indicate that aggregation of traditional microeconomic production functions into functions of the Cobb-Douglas or CES types is virtually impossible. Lave (1966, p. 141) points out that Samuelson's model, mentioned earlier, can be aggregated, but he doubts if its rigid assumptions permit its general use.

At this point, one must abandon the whole venture or reaffirm Solow's comment (1957) that some people like this sort of thing and others don't.

Usefulness of Empirical Studies

Nadiri, Walters, Lave, and others have suggested that many of the studies in this area have not really accomplished anything other than measuring the distribution of income in the areas studied, especially when Cobb-Douglas functions have been employed. Statistical significance of the results, they contend, is generated more by the well-behaved nature of aggregate data than by basic relationships among the productive agents.

Massell, as cited in Kennedy and Thirlwall (1972), reports that one-third of the reported advance in technology in such studies is not really technological advance at all, but merely reflects the movement of workers from more efficient industries to less efficient industries over time.

Denison (1962, 1967) and Jorgenson and Griliches (1962, 1966) have used the production function to organize the variables and then sought to make adjustments for quality and utilization to measure technical change. Denison found growth rates of technology in nine countries that ranged from 2.31 to 7.31 percent, per year, from 1950 to 1962. Jorgenson and Griliches used similar methods and concluded:

> if real product and real factor input are accurately accounted for, the observed growth in total factor productivity is negligible (1962, p. 249).

It would be easier to tolerate such inconsistencies if they weren't so frequent. Unfortunately, the literature abounds with such contradictions, giving some credence to Lave's suggestion that the estimates might behave as random numbers after being so thoroughly massaged by the computers.

Atkinson and Stiglitz (1969) have argued that the reported results do not and cannot measure technological change because such change is highly localized within the industries, and the aggregate data cannot capture this facet of the problem.

Perhaps the most damaging criticism of the overall approach discussed so far is one that is implied in Nadiri's 1970 review article. In introducing the technical part of his paper, Nadiri suggests the following as constituting the major unanswered questions in the area:

1. What determines the stock of pure knowledge in a society?
2. How and when does part of this knowledge take the form of innovations?
3. Which industries are likely to initiate adoption of the new techniques?
4. What are the characteristics of the transmission mechanism that underline the diffusion of new technology throughout an economy?
5. What are the external economics (diseconomies) of employing the new techniques?

By the end of his review, Nadiri has not used a single study that relies on an empirical production function or productivity ratio to help answer any of these questions.

Additional Criticisms

It is the opinion of this research group that the foregoing discussion casts so much doubt on the validity of studies of technological change oriented toward estimating empirical production functions that little is likely to be gained by devoting much more resources to each venture. Moreover, it may help round out this discussion to offer some additional criticisms of our own.

1. We believe that the complicated manipulation of crude data through multi stage estimation procedures accomplishes relatively little. To illustrate, we offer a simple example. It is widely acknowledged among growth theoreticians that in the absence of technological change, increasing output per worker over the long run is impossible. This is a fundamental conclusion of both Harrod-Domar and neoclassical models of growth. From the latter models, it is also argued that the rate of increase in output per worker will be equal to the rate of growth of technological change divided by the elasticity of output with respect to labor. The inside cover of a recent economic principles book (Spencer, 1971) shows that GNP in constant dollars increased from $355 billion in 1950 to $720 billion in 1970, or a rough doubling in twenty-one years. If we divide 0.69 (the natural logarithm of 2) by 21, we obtain 3.3 percent as the average exponential rate of growth for per capita output over the period. If we now multiply that figure by a rough estimate of the elasticity of output with respect to labor (0.75 is the most commonly cited figure), we obtain 2.48 percent as the average annual rate of growth of technology for the period. Although we do not believe that such a concept really exists, this estimate is close to what many "more sophisticated" techniques have produced, and it conforms to what all too often is the only criterion for judging such things—that is, "it does not seem unreasonable," although, of course, we have no meaningful criteria for judging "reasonableness."

2. We contend that any aggregative technique for estimating the advance of technology will inevitably mask significant changes in the component economic units and encourage misleading judgments concerning the effects of technological changes. "For example, since investment decisions are made at the level of the firm, and the employment effects of such decisions may differ widely among industries—depending on the relative pressures of labor-displacing technologies and of expanding product markets—the use of national aggregates to estimate the effects of adjustments in investment on employment levels may be of limited value" (Gold, 1971, p. 4). And the differential impact on industries of resource

stringencies and of shifts in product mix as well as of technological advances, combined with changes in the relative growth of industries with different input-output relationships, makes it equally hazardous to try to deduce the average rate of technological progress from analyses of national aggregates for outputs and inputs—or even to attach operational meaning to the concept.

3. Adjusting the labor input for variations in quality is not appropriate if the purpose is to recognize technological change, for many of the changes in the quality of the labor force are the consequences of changes in technology in prior periods. For example, increased educational attainments by workers are the result of technological advances that first freed man from the farms and later from the factories.[10] Aggregate techniques and production functions cannot detect such consequences.

4. Because of the feedback of technology into future inputs, there is a problem of simultaneity in production functions that has not been given significant recognition. A variety of other problems are also confronted in seeking congruence in the coverage of input and output measurements. (See Gold, 1965, pp. 290-91.)

5. The weights used in constructing dollar amounts of inputs and outputs are often inappropriate. For example, different outputs are often aggregated by weighting each output by its price; yet, marginal costs would be the appropriate weights if the study was intended to help optimize the utilization of society's resources—and in many industries, price is not equal to the marginal cost. Moreover, when output is measured in terms of value added, it is patently inaccurate to use final product prices as the weights (Gold, 1955, pp. 94-95).

6. Although the difficulties of measuring output have been cited by many authors, the criticisms on this point are still not complete. The electric power industry has been studied repeatedly because data are readily available and because the kilowatt hour supposedly represents a measurement of output undiluted by quality changes and product-mix variations. However, these studies all concentrate on the *generation* of electric power and seem to forget that the kilowatts do no good until they have been *transmitted* to the point of utilization. Enormous amounts of capital have been devoted to transmission of power that has not found its way into the production function. Furthermore, even if transmission had been considered, the "quality" of the kilowatt hour would still have been altered if probabilities of interruptions in service or brownouts had been affected. The moral is simple: there are many pitfalls in treating even the simplest output (or input) measures.[11]

7. We believe that insufficient progress has been made as yet in adjusting labor force data for the vast heterogeneity of and major changes in such inputs. Among other weaknesses, one glaring error in nearly all empirical studies is neglect of the growing tendency to have larger and larger percentages of total employees in salaried classes rather than as wage recipients. This tendency has decreased the flexibility of managers to adjust the labor force quickly in

response to changes in demand and other forces that affect the manpower needs of the firm. In effect, salaried workers not fully utilized are no different than capital that stands idle. Thus, the labor force should be adjusted for its rate of utilization just as capital is. Moreover, part of the widespread reductions in wage-earner requirements per unit of output have resulted from the shifting of their tasks to salaried employees.[12]

8. We believe that the difficulties of attempting to measure the net effect of embodied technological change are almost insurmountable. For example, a 1975 vintage automated drill press with one operator may be four times as productive as its 1945 counterpart, whereas the value or cost in real terms may be only twice the 1945 version, and the real value of the products of the 1975 press may be only twice that of the 1945 model. In the aggregate, no improvement in productivity will be measured, while in the drill press industry "apparent productivity" will have doubled.[13]

The previous points all refer to difficulties in employing the production function approach to the evaluation of measurement of technological change. It is possible that an ingenious researcher may find methods of overcoming these problems, but we have other reservations about the concept itself.

9. The omission of materials from the production function is excusable only if such materials exist in perfectly elastic supply and if they are not substitutable for capital or labor. It has been shown conclusively by Boylan (1975) that such an assumption in the blast furnace of the steel industry is unwarranted because the utilization of pellets in the furnaces virtually doubled their capacity. On the other hand, in the ore districts, enormous amounts of capital equipment were required to compensate for (or substitute for) deteriorating ore quality (Peirce, 1974). There are many other industries where one can envision similar types of substitutability. The adoption of pelletizing was required in the first place because materials (ore of a certain quality) were *not* available in perfectly elastic supply. It is apparent that the failure to treat materials satisfactorily is a major deficiency of the traditional approach.

10. Investigators employing the traditional approaches have failed to link their empirical findings to the actual behavior of real firms and industries. As an example, consider Wolkowitz (1973), who fitted a VES function to two-digit SIC industries in the United States. He concluded that the rate of neutral technological progress ranges from:

> a low of ten per cent, per year (Miscellaneous)[14] to a high of eight per cent, per year (Electrical Machinery). For adjusted data, the range is from one per cent, per year (Textile Products) to 16 per cent, per year (Fabricated Metals). The majority of estimates ranged from one per cent, per year, with a mean of three per cent for unadjusted data.

If ever results cried out for elaboration, these surely do. What could possibly cause a decrease in technology of 10 percent per year? Surely this can be explained by the way in which a variable is measured or the way in which industries are aggregated. What were some of the innovations that caused technological change in fabricated metals to grow at an annual rate of 16 percent? How did major innovations in the steel industry affect the measured rate of technological change in that industry? Here is a splendid opportunity to see if the data and methods employed are consistent with and revealing of actual practices in the industries. Alas, illumination will not be given on these points, and we receive instead:

> These results accord well with the estimates of product-augmenting technical change for the entire manufacturing sector, and on the two-digit level of aggregation they are most consistent with Lovell's estimates, although my estimates are on the average slightly greater (Wolkowitz, 1973, p. 635).

If this were only one rare case of the way in which the validity of results is assessed by users of these traditional methods, we could rest more comfortably, but unfortunately this is the typical case. Solow admitted not having a hypothesis in mind when he did his 1957 study, and the area is free of investigators who built models of change, analyzed the models, and developed from them testable hypotheses concerning the relationships among variables. Though statistical tests are commonly performed to see if the values of the coefficients differ from zero by a significant amount, such hypotheses were never really implied at the outset, and the validity of the study rests neither on the acceptance nor rejection of such a hypothesis.

11. The production function approach provides estimates that are useful to no one. Government policy-makers could not possibly base policy designed to foster innovation on estimated growth rates for technology because such rates do not reveal how, why, or where such growth occurred. If the "stock of technology" grew on the national level, it must have been because of specific new processes, products, techniques, machines, or ideas that were implemented in specific terms. Therefore, we must retreat to microeconomic data in the real world to see what forces engendered these changes and what consequences they wrought (Gold, 1971, pp. 14-18).

Managers also find no useful information in the findings of studies using the traditional approach. This group must manage complex industrial enterprises with emphasis on achieving maximum profitability, but estimates of average past changes in input-output relationships over an extended time at the level of industries or the nation cannot facilitate the performance of this function. Even if a study showed that a certain rate of growth of technology in a certain firm over a certain period was the direct consequence of a specific innovation made by the firm, the study would not be able to indicate whether the innovation was profitable. Again, we must examine the nature of the innovation itself and trace its successively broader and longer term consequences for input requirements, costs, and, beyond that, to adjustments in revenues and profits.[15]

The basic conclusion of our investigation is obvious: the state of the arts in measuring technological change by the use of aggregated production functions is not satisfactory. The underlying theory is weak, the data are not suitable to the purpose, the methods of estimation are not suited to the data, and the results are unimportant. Almost twenty years of research have been invested in this area, and the economics profession is reasonably entitled to expect a more rewarding outcome.

If criticism is to serve a valuable function, it must point the way toward the improvement of a given situation. Although our findings cause us to put no hope in the prospect of gathering useful information from empirical production functions, we cannot say the same for the basic thrust of the productivity ratio approach. The common deficiency in those ratios is that in the attempt to overcome the problems of interactions between physical inputs and outputs on the one hand and product and factor prices on the other, expedients have been employed which camouflage the specific effects which are important. Kendrick's method, for example, maintains constant factor prices (or "smoothed" ones) to avoid confronting the obvious fact that the value of the total output equals the value of the total inputs.[16] However, by so doing, he has created sources of variance in his index that are largely artificial and incapable of clear interpretation. The same criticism may be applied to the work of Davis, Farrell, and others.

An attempt is needed to utilize the full array of available engineering and economic information without artificial manipulation of the components. Consider a pair of examples from macroeconomics to illustrate the point. The national income accounts indicate that GNP is equal to consumption expenditures, gross investment expenditures, government expenditures, and net imports.

$$Y = C + I + G + E \tag{2.14}$$

Equation 2.14 is really an identity, and no amount of increased accuracy in measuring the variables will affect the form of this relationship. It would be possible, like Kendrick and others, to develop an arithmetic index to show the change in GNP as a function of the changes in the ratios of its components. After all, the arithmetic index begins with the identity that the value of total outputs equals the value of the total inputs:

$$Q = wL + iK \tag{2.15}$$

If we construct an arithmetic index for GNP, using the coefficients of the right hand variables as the weights, then our formula will be:

$$Y_2/Y_1 = B^*[(C_2/C_1) + (I_2/I_1) + (G_2/G_1) + (E_2/E_1)] \tag{2.16}$$

If appropriate values for 1950 and 1960 are substituted into the formula, the value of B^* turns out to be 0.24. However, no economic meaning can be attached to B^*, which is simply the number required to make the right side of the equation equal to the left. At this point, it is obvious that what we have done represents nonsense. Any student of economics could have told us at the outset of this exercise that we need not pursue this type of index because we already had a theory of the behavior of the right side variables in the consumption function and in the marginal efficiency of capital schedule. By adding these theories to the accounting system given above, we can develop a theory of the determination of gross national product. The end result is a richer and more illuminating product than any contrived index could ever be. The theory suggests to us (a) why income changes; (b) how it changes; and (c) what can be done to control the change.

Similarly, we might consider the multiplicative relationship concerning the money stock and its rate of turnover:

$$Y = MV \tag{2.17}$$

From this identity, we could develop the geometric index:

$$dY/Y = c^*(dM/M + dV/V) \tag{2.18}$$

If we sought to measure velocity through proxies, statistical artifacts, and other means (much as we estimate the capital stock for production functions), our measurements would surely have enough slippage to give C^* a value different from 1.00, yet surely no one would dare call such a number the rate of technological advance in the financial world. Again, the proper approach is to examine the behavior of the components of the identity and provide thereby a monetary theory of income determination, as the modern quantity theorists have done.

We suggest that these two examples point the proper path to be followed in developing a sensible, coherent theory for examining the effects of technological change. Instead of forcing the accounting data into awkward molds, we should analyze the figures in terms of the relationships that they actually depict. The proper approach is to define the accounting figures in terms of meaningful economic variables that underlie them and then provide theories of behavior for these support variables. The end result would be, not the ability to measure something called "the rate of change technology," but rather the total comprehension of how the physical and financial variables of the firm interact and their relationships over time. With such knowledge, we will have not only a picture of how specific changes in production technology affected such important variables as input requirement, factor proportions, unit costs, outputs and revenues, but in addition we will have an understanding of how many other types of changes (such as factor prices, production levels, and shifts in product mix) affect the performance of the plant and firm.

If a single method of analysis can allow us to understand the complete nature, causes, and consequences of technological change, it must contain at least three important sets of relationships. First, it must be able to link together the basic productive variables: output, man-hours, materials volume, capacity, fixed investment, and utilization. Second, it must tie together labor costs, materials cost, and capital costs. Third, it must coordinate the managerial control ratios that identify the major determinants of profitability to the firm. The productivity network of Gold seems to be the only current method that meets all of these criteria.[17]

Notes

1. For a more detailed review of differences between the analytical needs of academic microeconomists and of business decision-makers (and their staff advisors), see Gold (1971, pp. 1-18).

2. A more complete discussion of the mathematical properties of production functions can be found in many places, including Brown (1967).

3. Elaboration and empirical investigation of these properties can be seen in Chiang (1974).

4. Nadiri (1970) cites a slightly different version of this formula.

5. For further discussion of the role of such changes among the factors affecting input-output relationships, see Gold (1955, pp. 113-14).

6. For empirical evidence of the range of diffusion rates of major technological innovations, see, for example, Gold, Peirce, and Rosegger (1970, p. 220).

7. For a fuller discussion of the problems of measuring the contributions of capital to production and the provision of constructive suggestions for fulfilling specified analytical objectives, see Gold (1955, pp. 18-30). For constructive suggestions for separating the relative contributions of labor and capital to output, see Gold (1955, pp. 188-96).

8. For an extended discussion of these problems of measuring inputs and outputs, see Gold (1956, pp. 15-45, 91-101, 110-23); for a review of the additional problems of making input and output measures congruent in coverage, see Gold (1965, pp. 290-92). The often ignored problems of differentiating output measures from total inputs are discussed in Gold (1965, pp. 305-306).

9. Some further implications of this need to differentiate between causal relationships and the passive resultants of processes actively functioning at other levels of aggregation are discussed in Gold (1965, pp. 295-99, and 1971, pp. 3-4).

10. For a fuller review of the range of effects of technological changes on inputs, see Gold (1973, pp. 181-82).

11. For a review of other problems involved in output measurement, see Gold (1955, pp. 41-43, 92-97, and also 1973, pp. 182-85).

12. For a more comprehensive review of the changing role of salaried employees in industry and of the effects of such changes, see Gold (1955, pp. 155-64).

13. The problems involved in attempting such comparisons and their relevance for economic analyses and managerial decision-making are discussed in Gold (1955, pp. 18-30).

14. The imagination is titillated by an industry called "miscellaneous," and it seems foolish to think that something good could possibly come from this type of investigation of such an industry. However, it is no more foolish (or less so) than combining all industries into a national level of aggregation.

15. For such a broader analytical framework, see Gold (1965, pp. 293-95).

16. For a discussion of the problems of distinguishing measures of total outputs from measures of total inputs, see Gold (1965, pp. 305-306).

17. Originally published in 1955 (pp. 271-77), this approach was elaborated upon in 1956 and presented again with empirical applications in 1964, before undergoing further development in 1971 (pp. 19-31, 89-94, and 152-61). In England, some elements of a somewhat similar approach were brought to the attention of accountants in 1965 by Risk.

References

Abramowitz, M., "Resource and Output Trends in the United States since 1870," *American Economic Review* (May 1956).

Ames, E., "Research Invention, Development and Innovation," *American Economic Review* (June 1961).

Arrow, K., H. Chenery, B. Minhas, and R. Solow, "Capital-Labor Substitution and Economic Efficiency," *Review of Economics and Statistics* (August 1961).

Atkinson, A., and J. Stiglitz, "A New View of Technological Change," *Economic Journal* (September 1969).

Ball, R.J., "The Use of Value Added in Measuring Managerial Efficiency," *Business Ratios* (Summer 1968).

Beckmann, M.J., and R. Sato, "Aggregate Production Functions and Types of Technical Progress: A Statistical Analysis," *American Economic Review* (March 1969).

Boylan, M.G., *Economic Effects of Scale Increases in the Steel Industry* (New York: Praeger Publishers, 1975).

Brown, M. (ed.), *The Theory and Empirical Analysis of Production* (New York: Columbia University Press, 1967).

Bronfenbrenner, M., and P.H. Douglas, "Cross-Section Studies in the Cobb-Douglas Function," *Journal of Political Economy* (December 1939).

Brunner, E., "Some Shortcomings in the Economic Analysis of Technological Change," *Omega* (February 1974).

Chenery, H.B., "Engineering Production Functions," *Quarterly Journal of Economics* (February 1949).

Chiang, Alpha C., *Fundamental Methods of Mathematical Economics*, 2nd ed. (New York: McGraw-Hill, 1974).

Cobb, C.W., and P.H. Douglas, "A Theory of Production," *American Economic Review, Supplement* (March 1928).

Davis, Hiram, *Productivity Accounting* (Philadelphia: University of Pennsylvania Press, 1955).

Denison, E.F., *The Sources of Economic Growth in the U.S. and the Alternatives Before Us* (New York: Allen & Unwin, 1962).

_______, *Why Growth Rates Differ: Postwar Experience in Nine Countries* (Washington, D.C.: The Brookings Institution, 1967).

Domar, E.D., "On Total Productivity and All That," *Journal of Political Economy* (December 1962).

Eilon, S., and J. Teague, "On Measures of Productivity," *Omega* (October 1973).

Farrell, M.J., "The Measurement of Productive Efficiency," *Journal of the Royal Statistical Society*, series A, Part III (1957).

Ferguson, A.R., "An Airline Production Function," *Econometrica* (January 1951).

Ferguson, C.E., "Substitution, Technical Progress and Returns to Scale," *American Economic Review*, Papers (May 1965).

_______, *The Neoclassical Theory of Production and Distribution* (Cambridge: Cambridge University Press, 1969).

Friedman, Milton, "The Methodology of Positive Economics," in *Essays in Positive Economics* (Chicago: University of Chicago Press, 1953).

Galatin, M., *Economics of Scale and Technological Change in Thermal Power Generation* (Amsterdam: North-Holland, 1968).

Gold, Bela, *Foundations of Productivity Analysis* (Pittsburgh: University of Pittsburgh Press, 1955).

_______, "New Managerial Control Ratios," *Advanced Management* (April 1956).

_______, "Productivity Analysis and System Coherence," *Operational Research Quarterly* (September 1965).

_______, *Explorations in Managerial Economics: Productivity, Costs Technology and Growth* (London: Macmillan; New York: Basic Books, 1971).

_______, "The Impact of Technological Innovation—Concepts and Measurement," *Omega* (April 1973).

_______, and R.M. Kraus, "Integrating Physical with Financial Measures for Managerial Controls," *Journal of the Academy of Management* (June 1964).

_______, W.S. Peirce, and G. Rosegger, "Diffusion of Major Technological Innovations in U.S. Iron and Steel Manufacturing," *Journal of Industrial Economics* (July 1970).

Griliches, Z., "Research Expenditures, Education and Aggregate Agricultural Production Function," *American Economic Review* (December 1964).

Henderson, J.M., and R. E. Quandt, *Microeconomic Theory*, 2nd ed. (New York: McGraw-Hill, 1965).

Johnston, J., "Statistical Cost Functions in Electricity Supply," *Oxford Economic Papers*, n.s. (1952).

_______, *Statistical Cost Functions* (New York: McGraw-Hill, 1960).

Jorgenson, D., and F. Griliches, "The Explanation of Productivity Change," *Review of Economic Studies* (June 1962).

_______, "Sources of Measured Productivity Change: Capital Input," *American Economic Review* (May 1966).

Kaldor, N., "Comment on Professor Arrow's Paper, 'The Economic Implications of Learning by Doing,' " *Review of Economic Studies* (June 1962).

Kendrick, J., *Productivity Trends in the United States* (Princeton: Princeton University Press, 1961).

Kennedy, C., and A. Thirlwall, "Surveys in Applied Economics: Technical Progress," *Economic Journal* (March 1972).

Klotz, B., *Industry Productivity Projections: A Methodological Study* (U.S. Dept. of Labor, Bureau of Labor Statistics, 1966).

Lave, L.B., *Technological Change: Its Conception and Measurement* (Englewood Cliffs, N.J.: Prentice-Hall, 1966).

Lomax, K., "Cost Curves for Electricity Generation," *Economica* (May 1952).

Lovell, C., "Biased Technical Change and Factor Shares in United States Manufacturing," *Quarterly Review of Economics and Business* (Autumn 1969).

Maddala, G.S., "Productivity and Technological Change in the Bituminous Coal Industry, 1919-1954," *Journal of Political Economy* (August 1965).

Mansfield, E., *The Economics of Technological Change* (New York: Norton, 1968).

Moroney, J.R., "Cobb-Douglas Production Functions and Returns to Scale in U.S. Manufacturing Industry," *Western Economic Journal* (December 1967).

Murti, V., and V. Sastry, "Production Functions for Indian Industry," *Econometrica* (April 1957).

Nadiri, M.I., "Some Approaches to the Theory and Measurement of Factor Productivity: A Survey," *Journal of Economic Literature* (December 1970). Quoted with permission.

Nordin, J., "A Note on a Light Plant's Cost Curves," *Econometrica* (July 1947).

Peirce, W.S., "Factors Affecting Responsiveness to Technological Innovations in Coal Mining," Working Paper 54 (Research Program in Industrial Economics, Case Western Reserve University, 1974).

Risk, J.M., "Productivity Yardsticks," *Management Accounting,* (1965).

Robinson, J., "The Production Function, and the Theory of Capital," *Review of Economics Studies* (1953-54).

Sahgal, B., *Measures of Productivity and Efficiency–An Empirical Study*, MSc. Report, Imperial College, London, 1972.

Sahota, G., "The Sources of Measured Productivity Growth: United States Fertilizer Mineral Industries, 1936-1960," *Review of Economics and Statistics* (May 1966).

Samuelson, P., "Parable and Realism in Capital Theory: The Surrogate Production Function," *Review of Economic Studies* (1962).

Sherrard, W., "Labor Productivity for the Firm: A Case Study," *Quarterly Review of Economics and Business* (Spring 1967).

Solow, R., "Technological Change and the Aggregate Production Function," *Review of Economics and Statistics* (1957). Quoted with permission.

_______, "Investment and Technical Change," in K. Arrow, S. Karlin, and P. Suppes (eds.), *Mathematical Models in the Social Sciences* (Stanford: Stanford University Press, 1960).

Spencer, M., *Contemporary Economics* (New York: Worth, 1971).

Tinbergen, J., "Zur Theorie der lang tristigen Wirtschaftsentwicklung," *Weltwirschaftliches Archiv* (1942); ("On the Theory of Trend Movements," in *Selected Papers*, Amsterdam, 1959).

U.S. Department of Labor, Bureau of Labor Statistics, *Productivity: A Bibliography*, BLS Bulletin 1226 (1958).

_______, Bureau of Labor Statistics, *Productivity: A Bibliography* BLS Bulletin 1514 (1966).

_______, Bureau of Labor Statistics, *Productivity: A Bibliography*, BLS Bulletin 1776 (1973).

Walters, A.A., "Production and Cost Functions: An Econometric Survey," *Econometrica* (January/April 1963).

Winston, G.C., "Theory of Capital Utilization and Idleness," *Journal of Economic Literature* (December 1974).

Wolkowitz, B., "Estimation of a Set of Homothetic Production Functions: A Time Series Analysis of American Postwar Manufacturing," *Southern Economic Journal* (April 1973). Quoted with permission.

3 Productivity, Technological Change, and Input-Output Analysis

Asim Erdilek

This study seeks to appraise the applicability of the conceptual framework and measurements of input-output analysis to determining the effects of changes in productivity and technology.

Input-output analysis (I-O) is a versatile tool with many applications in theory as well as in practice. Its principal application is generally accepted to be the achieving of consistency in multisectoral resource allocation problems, such as those confronted in formulating action programs in business or government. This paper is concerned with productivity and technological change, however, an application in structural analysis that deals with interrelationships among the parameters of an economic model. Moreover, our analysis will be focused primarily on the plant, firm, and industry levels that are the immediate loci of such changes. Our discussion will also cover relevant applications of linear programming and activity analysis.[1]

Basic Objectives, Concepts, and Hypotheses

The Open Static Model

It has been suggested that we distinguish between input-output tables, "a vast collection of data describing our economic system," and input-output analysis, a "technique for explaining and predicting the behavior of our economic system" (Christ, 1955, p. 137). But Leontief has argued that "the compilation of an input-output table cannot be judged simply as a strictly descriptive enterprise but, on the contrary, must be interpreted as the fact-finding component of a rather elaborate analytical venture" (1955, p. 14). The primary analytical purpose of input-output is to achieve a consistent solution to a multisectoral resource allocation problem. Within the open static input-output model, this entails the determination of the sectoral gross output levels for a given set of sectoral final demands.

The open static model, which has had the widest empirical application, is basically a supply model based on a specific type of production function. It treats the demand side (including all final uses of sectoral outputs) as given and is thus unable to determine any variables on that side. On the supply side, it determines the sectoral production or gross output levels by making the following assumptions:

1. There is only one sector or process or technique to produce each good. In other words, there is no substitutability among intermediate inputs and/or primary factors in the production of any good, and all inputs are perfect complements.
2. There are no joint products; each sector or process or technique produces only one final product.
3. There are constant returns to scale in the production of each good.
4. Unless sectoral capacity and labor force constraints are exogenously given, they are assumed to be satisfied by the endogenously determined sectoral gross outputs.

Assumption 1 is the most controversial, followed by 3. They have been defended by input-output analysts on grounds of "practical necessity rather than theoretical convenience" (Leontief, 1955, p. 18). These analysts have suggested that the linear homogeneity and strict proportionality assumptions could be relaxed in empirical applications by piecewise-linear approximation of nonlinear relationships between inputs and outputs. Although substitutability among inputs has never been denied by input-output economists, it has been argued that substitution is not very important in many cases, especially in the short run. Clearly, this is not a general argument; its validity can be assessed only empirically. On the theoretical side, Klein (1952-53, pp. 131-36) has shown that the constancy of input-coefficients is consistent with joint products and a Cobb-Douglas type production function that permits substitution among inputs. But there are two important points to note in Klein's analysis: (a) the input-output coefficients whose constancy he has demonstrated are value, not physical, coefficients; and (b) these value coefficients lose their technological meaning and become dependent on the parameters of consumer demand and factor supply functions when the assumption of perfect competition underlying Klein's analysis is dropped.

In the linear programming formulations of the static open input-output model, the assumption that there is only one technique to produce each product is dropped, but the assumptions of no joint products and constant returns to scale are maintained. Thus, the linear programming formulation does permit substitution among inputs, and in doing so changes the basic nature of input-output analysis from that of a consistency model to that of an optimization model. However, the linear programming formulation of the static open model is not a full-fledged activity analysis model because it does not allow for joint products.

Assumption 2 creates an aggregation problem, referred to as the product-mix problem. Its seriousness decreases with increasing disaggregation, assuming 1 holds. If 1 does not hold, however, then we are also confronted with a process-mix problem that interacts with the product-mix problem, giving rise to a serious identification problem in empirical input-output analysis, which will be discussed below.

Assumption 3 can be dropped in two different ways: (a) capacity and labor force constraints can be explicitly incorporated into the static open model by the linear programming formulation, converting it from a mere consistency technique to a consistency-feasibility technique; and (b) the open model can be dynamized by taking investment out of the exogenous final demand vector and making it endogenous to determine sectoral capacity levels (but treating the labor force exogenously). Dynamic input-output analysis is more demanding than static input-output analysis in terms of both theoretical complexity and data requirements for empirical implementation. The theoretical complexities, especially the convergence problem, are still not well understood.[2] There are several dynamic input-output models based on alternative specifications of the investment function.[3] The empirical data requirements of the dynamic analysis are very hard to meet with available statistics. The capital requirements matrix is far more difficult to construct than the current requirements matrix. Furthermore, the former may dominate the latter in the process of productivity changes, which will not be clearly understood without the proper conceptual and statistical alignment of the two matrices.

Carter, one of the most prominent input-output analysts, has noted that the distinction between the endogenous and exogenous sectors in the static open model in terms of technologically determined structure (which only the former are assumed to possess) is not fully defensible (1970a, pp. 20-21). She points out that the exogenous—that is, the final demand—sectors are also technologically conditioned and that structural changes in the exogenous and endogenous sectors often interact. This interaction will be extremely important in the analysis of productivity changes, especially between the gross capital formation vector of final demand and the current intermediate input requirements.

Furthermore, it is difficult to make a sharp analytical distinction between current account items, which are put in intermediate input transactions, and capital account items, which are put in the gross capital formation vector of final demand. The classification of inputs into mutually exclusive flows (or current account items) and stocks (or capital account items) on the basis of their length of stay in the productive pipeline is arbitrary because the spectrum of length of stay in the pipeline is continuous (Carter, 1970a, p. 21). The treatment of inventories in input-output accounting is problematic because no clear cut distinction can be made between current account and capital account items and also because transactions data reveal sales and purchases, not production and productive consumption levels (Dorfman, 1954, p. 123). In constructing an input-output table, net inventory changes are shown separately as a final demand vector, but there are still questions as to how accurately they can be measured and, thus, how accurately the intermediate input-output coefficients can represent technological relationships.

The Concept of Productivity

Attention may now be turned to the concept of productivity in the static open model. In his classical book, *The Structure of American Economy, 1919-1939* (1951), Leontief first introduced the productivity concept in his input-output framework as follows:

Specify the general production function for industry i, $i = 1, \ldots, n$

$$X_i = F(x_{i1}, x_{i2}, x_{i3}, \ldots x_{ik}, \ldots x_{in})$$

as

$$x_{i1} = \alpha_{i1} X_i,\; x_{i2} = \alpha_{i2} X_i,\; \ldots, x_{ik} = \alpha_{ik} X_i, \ldots, x_{in} = \alpha_{in} X_i$$

where: X_i = the net physical output (total output minus the amount of its products consumed within the same industry) of industry i

x_{ik} = the amount of *kth* industry's products absorbed by industry i

$\alpha_{i1}, \alpha_{i2}, \ldots, \alpha_{ik}, \ldots, \alpha_{in}$ = the coefficients of production of industry i

To deal with the proportional variation in all the coefficients of production within any given industry, Leontief introduces common proportionality factors A_i, $i = 1, \ldots, n$, which he calls the *industry productivity coefficients* and writes (1951, p. 37):

$$x_{i1} = \frac{\alpha_{i1}}{A_i} X_i, \ldots, x_{ik} = \frac{\alpha_{ik}}{A_i} X_i, \ldots, x_{in} = \frac{\alpha_{in}}{A_i} X_i$$

initially setting $A_i = 1$ for all i, $i = 1, \ldots, n$.

In addition to A_i, Leontief introduces another set of proportionality factors, A'_k, $k = 1, \ldots, n$ and he calls A'_k the *productivity coefficient of commodity* k.[4] These coefficients measure the proportional change in the productivity of a given commodity k in all the consuming industries. (A_k coefficients measure proportional changes across a given column; A'_k coefficients measure proportional changes across a given row.)

Three remarks are in order regarding Leontief's productivity concepts. First, Leontief refers to changes in both industry and commodity productivities as "technical changes." Thus, it is impossible to distinguish between changes in productivity and changes in technology in his pathbreaking analysis. This lack of conceptual as well as empirical distinction between the two types of changes in input-output analysis has persisted until now. Second, his distinction between industry and commodity productivity concepts seems to be based on the interpretation of the columns of the input-output table as industries and of the rows of that table as commodities. This asymmetrical treatment of the columns and rows is usually regarded as essential to enhancing the technological significance of input-output analysis, as will be discussed in detail later. Third, based on his two productivity concepts, Leontief introduces a new concept—*the efficiency of a given commodity k* defined as the product of A_k and A'_k (1951, p. 64, p. 107).

> If A_k and A'_k were to change proportionately but in opposite directions, $A_kA'_k$ would remain the same. If the productivity of the coal industry had fallen to one-half its original level, but the combustion technique at the time had improved so as to reduce proportionally the amount of coal required in the production of any other commodity, the efficiency of the commodity coal considered from the viewpoint of the economic system as a whole would remain unchanged. The prices and quantities of all the other commodities would be the same, but the price of coal would be raised 100 per cent, and its output fall to one-half its previous level (1951, p. 64).

It is now quite clear that Leontief does not really make an explicit distinction between a commodity and an industry with the product mix problem in his mind, but merely distinguishes between a proportional change in the coefficients in the *column* of a given industry or commodity (change in "industry" productivity) and a proportional change in the coefficients in the *row* of that given industry or commodity (change in "commodity" productivity).

After the publication of the second edition of Leontief's *The Structure of the American Economy, 1919-1939*, Evans explored the conceptual relationship between labor productivity (the average product of labor) and technological change in the static open input-output model stressing interdependence among sectors (1953, pp. 33-53). After stating that technology determines labor productivity, he discusses the theoretical problems in the definition and the empirical difficulties in the measurement of labor productivity. We will review these problems and difficulties in a later section.

Evans, like Leontief, does not discuss the speed of productivity changes. His discussion of the sources of productivity changes is confined to those affecting *labor* productivity. Furthermore, he treats them exogenously and also does not show how they can be related specifically to the structure of the static open model.

In the construction of a labor productivity index, the index number problem arises from the product mix or the process mix problems or both. Evans argues that if labor productivity indexes are constructed for products or processes that are interdependent in production (such as final and intermediate products), then the resulting individual indexes and any of their averages will be misleading unless interdependence in production is taken into account. This is where input-output analysis comes in. It explicitly recognizes interdependence in production by focusing on transactions among production activities, thus considering the production and consumption of intermediate inputs that are ignored by the neoclassical production function approach. Evans points out that in the computation of total (direct plus indirect) labor requirements per unit of final demand, the basic interdependences in production are taken into account and that the reciprocals of these unit labor requirements are the proper labor productivities for different products or processes.[5]

Because its recognition of, and its ability to deal with, intermediate products is the chief strength of the input-output analysis, it seems appropriate to discuss why intermediate products should not be ignored. As Carter has argued (1970a, pp. 4-5):

1. Many (old and new) products in an industrialized, highly specialized economy are intermediate products.
2. Intermediate products are essential to bridging the gap between engineering data and conventional economic descriptions of production processes.
3. Consistent and detailed planning and forecasting by business and government are impossible without the explicit treatment of intermediate products.
4. Explicit introduction of intermediate inputs also contributes to the understanding of the traditional labor-capital-output relationships.
5. Technological changes which cause productivity adjustments affect industrial specialization (that is, interdependences among industries) as well as direct primary input (capital and labor) requirements of individual industries; and in fact, these two types of changes are interrelated.

Not all interdependences among production units are included explicitly in the standard input-output analysis. The interdependences that are explicitly recognized result from intermediate input transactions and the consumption of the same primary factors. These are market interdependences. Nonmarket interdependences—that is, external economies and diseconomies—are not taken into account by the standard static open model. It is possible, however, as Leontief (1970, pp. 262-271) has shown, to incorporate into the static open model interdependences in the form of external diseconomies such as arise from air pollution.

The first systematic discussion of how input-output could be used in structural analysis was provided by Leontief in 1953 and has been followed in later studies. Leontief (1953, p. 17) points out that "within the framework of an explicitly formulated theoretical system, economic change can be explained either as structural change or as a dynamic process." In structural analysis, exogenous changes in the parameters of the system provide the explanation for changes in the endogenous variables. This method of analysis is also called comparative statics. In dynamic analysis or dynamics, the laws of change, either constant or varying, are incorporated into the model, converting some if not all parameters of the static model into endogenous variables. Leontief's structural analysis is carried out within the framework of the static open input-output model.

He calls two economic systems structurally identical if their current flow intermediate input requirements matrices are identical (1953, p. 19). Structural change is defined as a change in the structural flow matrix. The causes or sources of structural change are taken as given. Leontief warns against the assumption of a nonexistent qualitative uniformity of goods and services produced by an economic system in its structural analysis over time. He does not, however, offer a solution to this aggregation problem. His adjustment of the different bench-mark structural flow matrices for a given economy for discrepancies in industrial classification and his reduction of those matrices to common physical units by appropriate deflation do not take care of the problem of changes in quality over time.

Leontief admits that a very serious limitation is imposed on his structural analysis by the fact that he focuses entirely on changes in the flow structure of the economy and ignores the changes in its capital structure (1953, p. 27). This limitation has already been noted in our earlier remarks concerning the relaxation of assumption 4 of the static open model.

According to Leontief (1953, p. 28), "a reduction in the input requirements per unit of output can be loosely described as an increase in productivity." This is different from his 1951 description only in the sense that he no longer insists on the proportional reduction in *all* the flow coefficients of a column or a row industry. He notes that in the column (input structure) of an industry, some flow coefficients would decline while others increase over time; that is, some inputs would become more, but others less, productive. Such opposite changes might reflect either technological change or merely substitution among different inputs due to changes in their relative prices. (However, such opposite changes might also be caused by changes in product mix.) It appears that Leontief's concept of productivity is very general in its coverage of the flow structure and is unable to say anything about the capital structure. However, one restriction is imposed on the coverage of the flow structure by the assumption "that the individual industries are structurally independent of each other in the sense that the technical *possibility* of substituting a new set of coefficients in any one column of a given structural matrix is in no way conditioned by the changes which might take place in any other column of the same matrix" (1953, p. 34). Leontief cautions us that the satisfaction of this assumption empirically depends on the industrial classification adopted. It is questionable, however, whether this assumption's empirical validity is only a matter of industrial classification.

Leontief describes labor productivity on an economy-wide basis. "An increase in the total number of man-hours required to produce a given bill of goods signifies a reduction in the productivity of labor, while a fall in labor requirement indicates that manpower has become more productive" (1953, p.

35). Like Evans, he argues that the proper measurement of labor productivity must take into account interdependences among industries that require each other's outputs as intermediate inputs. Therefore, the labor productivity for an individual industry or for the entire economy is the reciprocal of the relevant *total* (direct plus indirect) unit labor requirement. It is thus possible for labor productivity to change in a given industry although no input coefficient in that industry's column (including its direct labor requirement coefficient) changes. It is also possible for economy-wide labor productivity to change, although no individual industry direct labor requirement coefficient changes. It should be noted, however, that the changes in the economy-wide labor productivity are relative to a given bill of goods. This fact creates an index number problem in empirical analysis, which will be discussed later.

Carter, in her empirical work on the structural changes in the United States economy between 1939 and 1961, updating Leontief's own empirical work covering the period between 1919 and 1939, has followed the methodology formulated by Leontief. She uses the term "structural change" synonymously with the term "technological change," adopting the latter's definition by economists as "any change in the parameters of the particular form of production function they have chosen (1970a, p. 11). Any change in any of the intermediate input coefficients or in the primary factor (labor and capital) coefficients, is a structural or technological change in the static open input-output model. However, Carter finds this definition of technological change too restrictive and repeatedly, but peripherally, mentions other dimensions and forms of technological change that she cannot consider "within the self-imposed limits of an open static model" (1970a, p. 131). She seems to confine the term productivity to the two primary factors only. The effects of a technological change in terms of changes in the primary factor total requirements per unit of final demand are interpreted as changes in productivity. Like Leontief and Evans, she stresses "the importance of shifting industrial specialization in the changing productivity picture" (1970a, p. 133). By "shifting industrial specialization," she refers to the varying patterns of intermediate input transactions among different industries, or technological changes.

Although the static open input-output model cannot explain and relate changes in intermediate input, labor, and capital requirements over time, Carter hypothesizes and empirically confirms that "changes in all three are closely interwoven" (1970a, p. 135). She analyzes them within a "network of structural changes" that is exogenous to the static model. This network is not a formal model by itself; it is simply the hypothesis that changes in intermediate input, labor, and capital coefficients are interrelated.

Shortcomings of the Treatment of Structural Change

Two basic conclusions seem to emerge from our review so far. First, in input-output analysis, as in neoclassical production theory, there is no clear-cut distinction between technological change and productivity adjustment.[6] Second, in input-output analysis, the theoretical and empirical study of productivity adjustment is conducted within a comparative static framework. This means that the sources of productivity adjustments are taken as given, and that the speed of such adjustments cannot be explained endogenously. In this respect, too, input-output analysis and neoclassical production theory are similarly limited in their scope. The static open input-output model cannot explain the timing of productivity adjustments. To be able to do so, it must use a full-fledged dynamic model in which structural change is endogenous. No such model has been constructed to date, or at least has not been empirically implemented.

In his comments on a classic paper by H.A. Simon, Brozen has raised the question whether technological change could be treated endogenously in a linear activity analysis model (1951, pp. 279-281). Brozen notes that it will not do to make the technological coefficients for different activities a function of time, depending on (a) research and development, (b) innovation, and (c) diffusion. Can (a), (b), and (c) be introduced into the technology matrix as separate activities (as additional columns)? Yes, but then the technological coefficients become functions of previous levels of these three activities, creating nonlinearities. To avoid nonlinearities, suggests Brozen, these three activities must be introduced as rows, as well as columns, and their outputs must be treated as inputs to other activities. Now if (a), (b), and (c) can be placed within the technology matrix, the question of their optimum levels arises. But to answer that question, we need empirical data for the rows and the columns of (a), (b), and (c). Provided such data can be obtained, a linear programming model can be set up to find the optimum levels of R&D, innovation, and diffusion, concludes Brozen. As Kamien and Schwartz (1975, pp. 1-37) and Kennedy and Thirlwall (1972, pp. 11-72) argue, however, such data are not readily available, and the hypotheses regarding the determinants of R&D, innovation, and diffusion are themselves highly vague and controversial. Even if such data were available, Brozen's suggested formulation is inadequate because it cannot yield an explicit and dynamic treatment of endogenous technological change and productivity adjustment. The contributions of (a), (b), and (c) to other productive activities are stated in terms of current input requirements of the latter, and not in terms of *changes* in their current input requirements. However, it is doubtful whether technological and productivity adjustments can be explained adequately even in a dynamic model if it is only a production model—that is, a model which takes no cognizance of changes in tastes, factor prices, market imperfections, and environmental conditions.[7]

Analytical Foci

This section will analyze the coverage and measurement of inputs and outputs in input-output analysis, including the index number and the aggregation problems mentioned earlier.

The Coverage and Classification of Inputs

Input-output analysis covers both primary and intermediate inputs. In the static open model, both labor and aggregate capital—that is, the entire fixed capital stock consisting of buildings and machinery—are treated as primary factors. Labor may be disaggregated into several skill or occupational categories, but capital is regarded as a homogeneous input.[8] In a dynamic input-output model, capital becomes an intermediate input disaggregated by both producing and consuming sectors or industries. The capital matrix contains these capital stock intermediate input requirements. So, when we go from static to dynamic input-output analysis, capital is converted from a homogeneous primary factor to a set of heterogeneous intermediate inputs.

In the terminology of input-output analysis, direct inputs and indirect inputs have different meanings from the same concepts in production or cost accounting. In input-output analysis, a direct input is any (primary or intermediate) input that a sector or process consumes directly—such as all the labor used in a plant, regardless of its job classification (production, maintenance, service, or administration)—is a *direct input.* An *indirect input* is an input consumed indirectly through intermediate inputs embodying that input—for example, the automobile sector may or may not consume coal as a direct input (to heat its buildings), but it definitely consumes coal indirectly through its direct consumption of steel that consumes coal directly. The direct input requirements are expressed in terms of the direct technological coefficients of the current flow matrix, labor vector or matrix, and capital stock vector or matrix. These requirements are expressed with respect to sectoral gross output levels. The indirect input requirements are combined with the direct input requirements in the so-called total (direct plus indirect) requirements. These total requirements are expressed with respect to sectoral final demand levels and in terms of the coefficients of vectors or matrices derived from the inverses of the direct requirements-related matrices.

In the formulation of direct current flow intermediate requirements, a sector's own consumption of some of its output may be either netted out from its gross output, or shown separately. The first alternative yields an **A** matrix with a zero principal diagonal, whereas the second and much more common alternative may yield an **A** matrix with a nonzero principal diagonal. The second alternative, then, covers both purchased and internally produced intermediate inputs.

As a final note concerning the coverage of inputs in input-output analysis, there is a serious conceptual and empirical difficulty in making a sharp distinction between current flow and capital stock intermediate requirements, as discussed in the previous section. Input-output analysis is capable of covering all inputs, but its distinction between inputs required for current operations with a fixed productive capacity and inputs required for enlargement of productive capacity is ambiguous. This problem is, of course, camouflaged by static input-output analysis to some extent, because it is vague on the question of productive capacities. We should point out that the so-called capital flow matrix is not identical with the capital stock matrix of dynamic analysis, although they are related. The total interindustry transactions may be broken down into two categories for a benchmark year: (a) transactions involving noncapital goods transactions; and (b) transactions involving capital goods. The second category may then be expressed in terms of the capital flow matrix that pertains to the gross, not net, investment processes of different sectors. Of course, if we have a capital consumption or depreciation matrix, then by subtracting it from the capital flow matrix, we can find the capital transactions for net investment, from which we can derive the capital stock matrix of dynamic analysis. The capital goods transactions for depreciation are then added to noncapital goods transactions to obtain the current flow requirements of static analysis.

Measurement of Inputs

Attention now turns to the measurement of inputs in input-output analysis. The first question that requires an answer is whether inputs should be measured in physical or value units. In *The Structure of American Economy, 1919-1939*, Leontief's initial formulation is in physical units for both inputs and outputs. He then introduces prices explicitly to convert the model in physical units into the model in value units.

Later, however, Leontief dispenses with the separate price and quantity figures and conducts his empirical analysis largely in terms of value relationships alone. After expressing all intermediate input transactions in dollar figures, he notes that "in order to obtain the corresponding physical amounts of all commodities and services, we simply define the unit of physical measurement of every particular type of product so as to make it equal to that amount of the commodity which can be purchased for one dollar at prevailing prices" (1951, p. 72). The "dollar's worth" is a meaningful physical unit, however, only if prices stay constant.

In the introduction to his empirical structural (productivity) analysis, Leontief (1951, p. 109) points out that:

> The change in any efficiency coefficient $A_k A'_k$ caused by a variation in A_k alone affects the *value* setup of the system exactly in the same way as an equal change in A'_k alone, or as any equivalent combination of the two. To avoid unnecessary complexity in the following formal analysis, the productivity coefficients A'_k of the various commodities will be considered as constant and all the variations in the corresponding efficiency coefficients $A_k A'_k$ will be entirely attributed to the industrial coefficients A_k.

In short, because Leontief's structural analysis—that is, investigation of the intermediate flow matrix changes—is based on *value* data only, he cannot express the changes in value coefficients in terms that *separate* physical input and price changes.

Carter's structural analysis is also based on value data. She notes that the price and quantity components must be separated to relate structural change to technological change (1970a, p. 22). Because "the dollar's worth" as a physical unit changes as prices change over time, the comparison of the input-output matrices of different years within the static open model can be made technologically meaningful by deflating all the value coefficients of the different years to a common price basis.

Carter admits that this deflation process is both conceptually and empirically difficult. First, there is the index number problem—which year is to be chosen as the base year. Second, there is the quality change problem. Third, in the specific case of input-output, there is no unique deflation procedure; that is, deflation can be cell by cell, by row, by column, or by any combination thereof.

Of course input-output data *can* be collected and processed in physical units; and, consequently, input-output analysis *can* be conducted in those physical units. This can be accomplished, however, only on a very disaggregated level. It is not feasible on the national level, the level at which the structural analyses of both Leontief and Carter have been conducted. The circumstances under which it is feasible to conduct input-output analysis in physical units are limited to the uncommon cases of plants with stable product mixes and of industries consisting only of plants and firms meeting these requirements. As pointed out at the beginning of this survey, our interest in the actual and potential achievements of input-output analysis in the study of productivity adjustments is primarily at the microlevel—the plant, the firm, and the industry levels—because it is at these levels that key decisions are made and later evaluated concerning changes in products, processes, inputs, output levels, operation conditions, and other factors affecting productivity, technology, costs, and revenues. Unfortunately, we are aware of very few empirical studies that use input-output analysis at the microlevel to investigate productivity changes. These studies will be reviewed in the next section along with several theoretical studies, especially by cost accountants, that explore the possibilities of applying input-output analysis to management problems, including productivity changes.

If the input-output transactions are to be measured in monetary units, then the question arises whether purchasers' or producers' prices should be used—that is, how the distribution charges should be handled. Clearly, the technological significance of input-output coefficients would be enhanced by the use of producers' prices.

Our discussion of the measurement of inputs has been confined thus far to intermediate inputs. What about the measurement of labor and capital, the two primary factors of the static open model? Labor is usually measured in physical units; for example, man-years. Its treatment as a homogeneous input is open to objections based on qualitative differences among workers, especially in terms of training skills and experience. Such objections could be met by disaggregating labor into relevant analytical categories if the detailed data were available. As was pointed out earlier, capital is treated as a homogeneous input and measured in monetary units in the static input-output model. Of course, this capital is not homogeneous; it is simply the sum of the values of different types of buildings and machines. As such, its value is sensitive to price changes; attempts to measure it in constant prices give rise to an index number problem. There is, of course, also the problem of changes in quality of different types of physical capital over time involving deterioration of capital goods through use, which may or may not be correctly reflected by depreciation allowances, as well as technological improvements in later acquisitions, which may or may not be adequately reflected in price changes.[9] In dynamic input-output analysis, capital becomes a set of heterogeneous intermediate inputs which may be measured, like current flow intermediate inputs, either in monetary or physical units. Its measurement in physical units is feasible, however, only when intermediate inputs in general can be measured in physical units, which is more difficult to achieve in dynamic analysis than in static analysis because changes in capacity are very often accompanied by changes in product mixes and technologies.

Disaggregation of Inputs

The preceding discussion of the coverage and measurement of inputs (and outputs) in input-output analysis also touched on the question of their disaggregation. To be more explicit, in input-output analysis disaggregation of inputs is not limited conceptually except for capital in static analysis. Intermediate inputs can be disaggregated, by increasing the number of production sectors, as much as available data and data-processing capacity allow. Such disaggregation can permit distinctions among materials by kinds and quality grades relevant to substitution; for example, very fine disaggregation can allow separate treatment of different grades of iron ore in pig iron manufacturing. In dynamic analysis, capital intermediate inputs can also be disaggregated with no conceptual limits. In static analysis, however, capital is treated as a homogeneous input, with no distinction among different types of capital goods with respect to sector of origin or function. This is, of course, a very serious limitation which is also present in the neoclassical production function theory. Labor can be disaggregated in either static or dynamic models, again without any conceptual limits, with respect to several categories relevant to the purposes of the analyst.[10]

It may be asked if labor can be disaggregated, why can capital not be disaggregated in static input-output analysis? Simply because capital needs to be disaggregated by producing sectors, and if such disaggregation were possible, then dynamic analysis would become feasible and appropriate. Of course, each sector's aggregate capital stock could be disaggregated by a criterion other than producing sector—for example, domestically produced versus imported, or ten or more years old versus less than ten years old. Such criteria are not, however, of general interest, not being technologically very meaningful. But, yes, in terms of splitting a single aggregate capital stock restraint in a linear programming formulation of the static input-output model into several capital stock constraints, the disaggregation of capital as a primary input is certainly possible and, in specific cases, meaningful.

Coverage, Classification, and Measurement of Outputs

In reviewing the coverage of outputs, the earlier distinction should be recalled between gross output of a production sector and its contribution to final demand. The difference is that sector's contribution to demand for intermediate inputs. A distinction is also made between gross total output and gross domestic ouput of a sector. The difference between the two is that sector's imported output. The intermediate transactions may or may not include imports. If they do not, then there is no difference between gross total outputs and gross domestic outputs. Input-output coefficients derived from intermediate transactions of net imports are not technologically very meaningful. If intermediate transactions, as is often the case, do include imports, then the input-output coefficients are derived by dividing the flows in a sector's column by that sector's gross domestic output.

A sector's contribution to final demand includes its contribution to net inventory change. Consequently, a sector's output in input-output analysis is based on production, not sales.[11] A sector's gross domestic output is equal to its total revenue (also its total cost of production), if monetary units are used in measurement. Total revenue includes the total cost of intermediate inputs plus the value added, which includes wage (and salary) cost, profit, depreciation, net indirect taxes, and noncompetitive imports. The sum of wage cost, profit and depreciation is the cost of primary inputs. The sum of profit (all nonwage compensation) and depreciation is the share of capital in value-added or net output.

Because in input-output analysis, each production sector has dual roles as a producer *and* as a consumer of intermediate inputs, any disaggregation of intermediate inputs entails an identical disaggregation of outputs. In other words, because the number of producing (row) sectors is the same as the number of consuming (column) sectors (that is, the transactions matrices are square), standard input-output analysis makes no distinction between disaggregation of outputs and of intermediate inputs. This follows from the assumptions that each sector produces a single product and that each product is produced by only one sector. These assumptions rule out the product mix and process mix problems. But in many empirical applications of input-output analysis, these two problems confront the analyst and thus have to be dealt with in one way or another.

Product Mix and Process Mix Problems

Before we examine problems of product and process mix, let us again note that the deflation of value data that reflect current prices is mandatory in input-output analysis for any meaningful structural analysis. This deflation process is a particularly complex and laborious one empirically. The product mix and process mix problems compound the difficulties of deflating the value data of different benchmark years to make them comparable in common constant prices. These two problems may affect different sectors differently, necessitating different deflation procedures.

In his discussion of the product mix question, Leontief (1955, p. 21) has stated that "not only through its empirical setting, but also in terms of its theoretical formulation and solution, this question is very closely related to the explanation of technological change." In other words, changes in average input-output coefficients over time may be partly caused by shifts in the product mixes of individual sectors. Leontief, making a clear distinction between the product mix problem and the process mix problem, suggests that in some cases the former is caused by an inappropriate statistical classification. For process mix, which he treats as technological change, Leontief distinguishes between: (a) a change in the process mix of a sector reflecting shifts in its flow inputs structure only; and (b) a change in the process mix of a sector reflecting shifts in both flow and stock (capital) input structures of that sector. He considers (b) to be a more common occurrence than (a). Leontief does not deal with the question of distinguishing empirically between the problems of product mix and process mix. His distinction is conceptual only.

Like Leontief, Carter approaches the product and process mix problems from a macro—that is, national economy—viewpoint, reflecting the restriction of her empirical analysis of structural change to that level. She notes (1970a, p. 8) that a column of a national input-output coefficients matrix represents an average technique or process derived from data on the values of inputs and outputs for all the various establishments in a given industry: "No matter how detailed the classification, each output category will subsume a mix of products. . . . Even within a given establishment, many different products may be produced, each from a different combination of inputs." In other words, there is, first, the problem of product heterogeneity for different producers in the same industry; and, second, there is the problem of multiple products for individual producers in a given industry. The first problem implies the lack of perfect competition, and the second problem implies the violation of the assumption made by input-output analysis that each sector produces a single product. The problem of product differentiation defies any realistic empirical solution. The second problem is tackled empirically by separating the "primary product" of a given industry from its "secondary products," and treating the transactions involving the latter through fictitious transfers among different sectors.

Carter points out that the distinction between the changes in product mix and changes in process mix *within a particular sector* is forced by the conception of a sector as an aggregate, which, in turn, is forced by the restriction of the analysis to the macro level. This distinction is conceptually clear, although empirically it is not. But even conceptually, the distinction is not clearcut *from the viewpoint of the entire economy*, since often a change in the product mix of a particular sector is induced by changes in the process mixes of other sectors.

The process mix problem may arise either on the sector level or on the establishment level. On the sector level, the market shares of different establishments each using a single process to produce a homogeneous product may shift, causing a change in the process mix of the sector. On the establishment level, the mix of processes used to produce a single homogeneous product may change because of diffusion of new techniques. Raw steel, for example, is a relatively homogeneous product that can be produced with quite different input combinations by the basic oxygen process, the open-hearth process, and the electric furnace process. Process mix changes on the sector level will often reflect both changes in the market shares of different establishments and changes in the process mixes of the establishments themselves. A process mix change may also take place among processes that are already in use or may involve the introduction of new or possibly old processes that are not already in use. But because the basic data for input-output coefficients are obtained from detailed census statistics and because the census statistics cover only *actual* inputs and outputs, any given benchmark year input-output coefficients are related only to the mix of processes actually in use during that year.

The Aggregation Criteria

As we have said, the basic data for input-output technological coefficients are provided by census statistics. In arranging these statistics in terms of a two-way classification of inputs and outputs meaningful for input-output analysis, we need aggregation criteria.

Leontief, concerned with the classification of national census statistics collected on an establishment basis into an interindustry framework, stated (1951, p. 20) that ideally "all production enterprises should be segregated into several homogeneous industrial groups, homogeneity being defined in terms of (a) identity of products and (b) qualitative and quantitative similarity of the cost structure of the firms within each group." These are not, however, operational criteria. Leontief's remarks about the inevitability of "a compromise between a theoretical ideal and practical necessity" in actual statistical work provides no specific guidelines for aggregation, although they reveal the complexities encountered in the real world. His two specific criteria, which ensure the invariance of the price and the output derivatives of sectors not involved in aggregation, on the other hand, are too restrictive because (a) they assume that an input-output system of *n* sectors has already been constructed, and (b) they are themselves applicable only to very special cases. These two specific criteria are: "(a) neither of the two consolidated industries uses the product of the other and, at the same time, their output distributions and their cost structures are proportionately equal; (or), (b) the outlays of one of the two industries are devoted exclusively to the purchase of the products of the other, while at the same time the other industry does not buy any of its products" (1951, p. 127).

For their discussion of aggregation criteria, Evans and Hoffenberg are also concerned with the classification of national census statistics collected on an establishment basis into an interindustry framework. They define an industry "as a collection of productive units engaged in supplying commodities or services that are similar in nature and for which some common output measure is available," adding that "the similarity in outputs is expected to enjoin a certain degree of similarity in production facilities and in input requirements for the producing units within a sector" (1953, p. 74). They suggest that departures from this expectation can be remedied by the availability of maximum commodity detail on both input and output sides within sectors.

Their first principle of aggregation is "the principle of exclusive use." Two sectors may be aggregated without loss of information, if one of the sectors provides all or almost all of its output as an input to the other sector—for example, almost all of the output of iron ore mining is consumed by blast furnaces. Their second principle is "the principle of identical input structures." If the input-output coefficients of two sectors are identical or very similar, nothing is sacrificed analytically by their combination. Their third principle is "demand complementarity." This suggests the combination of two or more sectors if their products are complements for specific uses. This principle lacks general applicability. Their fourth and most general principle is "the principle of irrelevance." An input-output analyst studying a certain problem, such as the iron and steel industry's productivity adjustments, may combine those sectors, such as food and tobacco, which have no direct relevance to his problem (1953, p. 75).

Fisher, at the beginning of his analytical discussion of the aggregation question of input-output analysis (1958, p. 250), defines and answers that question as follows:

Given a detailed input-output table in which many "small" industries appear, and given the desire of the research worker to have only a few "large" industries to deal with, when is it possible to consolidate the small industries, and still obtain the same results of the analysis or predictions as would have obtained by using the many small industries in the first place? The answer to this question has been shown to be "almost never."

He notes that although the necessity for formulating "good" aggregation criteria had been accepted, no such criteria have been developed in operational terms. ("Good" aggregation criteria are those procedures which yield "approximately the same" results as yielded by the original model.) He develops such criteria, which he calls "minimal distance criteria," by using statistical theory (1958, pp. 254-256). These criteria help the analyst choose the sectors of the detailed input-output table that are to be consolidated on the assumption that the level of aggregation and the method of aggregating given sectors are predetermined. Now, it is quite obvious from what has been said above that Fisher does not concern himself with questions regarding the construction of the detailed original model. He takes it as given. Furthermore, his criterion for "good" aggregation criteria is based on a specific, although the most frequent, objective of input-output analysis: the determination of the sectoral gross output levels required for a given bill of goods—that is, a desired or projected vector of final demand. Fisher acknowledges the restrictive nature of this "accuracy of predictions of output" objective and implies that, for the structural analysis objective, his minimum distance criteria are not likely to be appropriate (1958, p. 260). Fisher's analysis is based on the static open input-output model in which transactions are measured in monetary units. Therefore, aggregation involves the summation of money flows in prices of a given year.

Leontief, in one of his later writings, adopts the objective of structural analysis as his objective and develops an analytical technique, which he calls "double inversion," as an alternative to aggregation. Like aggregation, the double inversion method is used to decrease the number of sectors in the original model. "It is, however, a 'clean'—not an index number-operation. It does not involve introduction of weights or any other arbitrary constants" (1967, p. 414). Leontief's objective is to make two input-output matrices for the same economy at two different points in time, or for two different economies, comparable for structural analysis. He finds aggregation unsuitable for this purpose since it provides comparability "at the cost of analytical sharpness in the description of the underlying structural relationships" (1967, p. 412). As an alternative, the double inversion procedure does not distort any of the basic structural relationships, but the comparability it provides is confined to those sectors of the two matrices which are directly comparable—that is, after double inversion both matrices have only directly comparable or common sectors. The incomparable sectors are reduced or analytically eliminated by incorporating their structural relationships into the comparable sectors.

Like Fisher, Leontief works within the static open model, but, unlike

Fisher, he operates on the input-output coefficients (for both the intermediate inputs and primary factors) directly instead of the monetary flows. The double inversion refers to two coefficient matrix inversions (1967, p. 414). Consequently, Leontief's method can be applied to coefficient matrices based on physical units of measurement for inputs and outputs, but Fisher's method cannot. Leontief does not note this difference because he, too, like Fisher, is concerned with national input-output tables that are based on monetary units.

To sum up, their analyses have almost exclusively focused on the proper procedures for consolidating a given input-output table or matrix instead of dealing with the question of how to arrange a set of census statistics into an input-output table by a two-way classification of inputs and outputs. Also, the aggregation problem and criteria have been discussed only on the national or macro level.[12]

Industry and the Product Approaches to Specification of Technology

The question of how to arrange census statistics collected on an establishment basis into an input-output table by a two-way classification of inputs and outputs with technological realism is raised and answered by Gigantes (1970, p. 270). His contribution may be viewed as an attempt to bridge the gap between input-output analysis and activity analysis in a limited sense: recognize explicitly the existence of by-products and joint products in input-output analysis and develop transformations between industry-based and commodity-based technologies to deal with such products systematically. Gigantes argues that this effort would increase the realism of input-output analysis and lend itself to empirical implementation, thanks to the capabilities of modern electronic data processing. He considers industry-based and commodity-based technology models as extremes and suggests that mixed-technology models would be needed in most applications.

Gigantes's analysis aims for a generalization of the open static input-output model by relaxing its rigid assumptions about the product structure of industries. His distinctions among industry-technology, commodity-technology, and mixed-technology models enables him to fit into the input-output framework *formally* diverse technological situations. He confines his analysis to the macro level but it can be applied to such micro production units as the firm or the individual plant. An industry can be split into several firms, leaving the number of commodities unchanged. The market-shares hypothesis can be applied just as well to firms. Again, formally, individual plants can be incorporated into a firm-industry model. There appear to be serious difficulties, however, in the application of Gigantes's analysis through actual computations, such as the appearance of negative elements in the commodity-technology matrix. Gigantes asserts that this problem is often caused by the "technological heterogeneity of industries." He then writes (1970, p. 281):

> Countries that conduct surveys of industrial establishments involving detailed questions on the inputs and outputs of these establishments can mitigate this deficiency in their industrial classification. For the purpose of calculating commodity input structures, establishments can be grouped with the criterion of technological homogeneity in mind. These groups of establishments, which will constitute new industries, must be chosen to produce products within a narrow range of commodity classification. This in turn implies that the commodity classification of inputs and outputs must be, so far as possible, technologically homogeneous. If these conditions are met the transformation involved in calculating the commodity technology matrix can be applied to the input structures of the new industries. The size of the resulting commodity technology matrix will depend on the number of new industries defined. In general, the criterion of technological homogeneity requires very large matrices.

The difficulty here, of course, is the ambiguity of "technological homogeneity." The noncontroversial last sentence in the quotation above does not directly specify what technology homogeneity is; it simply states the obvious fact that the treatment of diversity in reality requires detailed data. Furthermore, "industries" are synthetic; that is, they are created by the empirical analyst. Firms, however, are not; they are real-world entities that cannot be decomposed and regrouped along the lines suggested by Gigantes to reduce their technological heterogeneity.[13] Therefore, the application of his analysis to firms (and plants) cannot be assured of meaningful results by reclassification of transactions data. In short, Gigantes's analysis lacks general empirical applicability because of (a) ill-specified computational procedures and (b) unrealistic data requirements. Carter, in her structural analysis of the U.S. economy, mentions but does not employ methods such as that of Gigantes to estimate "pure" product-to-product input structures from the establishment-based census data. However, Carter feels that "the advantage of doing so for future work is apparent" (1970a, p. 9) to solve the product mix problem of conventional input-output analysis.

Applicability of Input-Output to Micro Production Units

For more than twenty years, both theoretical and empirical input-output analyses by economists have been almost exclusively on the macro levels, national and international, with regional applications increasing during the last decade or so. Although several economists have recognized and shown the applicability of input-output to such micro production units as the firm and the plant, they have until now neglected their systematic investigation by input-output analysis.

Gols, a business consultant economist who is one of the most knowledgeable specialists in the micro applications of input-output, reaches the following conclusion at the end of his review (1975, p. 27) of the potential uses of input-output in industry:

> perhaps the key stumbling block to input-output use in corporate planning is that the assumptions and definitions concerning technological change, structural change, pricing and demand sensitivity, on which the input-output coefficient changes are based, are of very great interest to the planner and of perhaps least interest to the input-output analyst; in fact, most input-output analysts tend to sweep these assumptions under the analytical rug, because they interfere with the mathematical elegancies of the input-output model or the simplicity of the analysis task.[14]

Gols is correct in pointing out that the micro applications of input-output require more realistic and flexible assumptions, especially with respect to technology, and that structural analysis must occupy the center of the stage in such applications. The real problem is not so much that economists have ignored or neglected structural analysis, it is that their structural analysis has been *ex post* or descriptive in nature, with little or no effort and success in providing endogeneous explanations of coefficient changes (and productivity adjustments). Such a structural analysis may be of interest and may not appear to be too limited in scope on the macro level. After all, at least in the United States, no national policy makes serious use of input-output and, therefore, requires no operational understanding of how and why structural change occurs on the national level.[15] Such an *ex post* structural analysis is of very little use to a policy maker like the corporate planner, who must know and understand the dynamics of structural change because it is one of his primary responsibilities to induce structural changes that will promote his objectives.

A Survey of the Relevant Accounting Literature

Attention turns next to the efforts of some accountants to apply input-output analysis to firm and plant management problems. In an early attempt, Richards presents a direct translation of Swift and Company's accounting system (using actual financial data for the 1951-1957 period) into the open static input-output model (1960, pp. 429-436). His application of input-output analysis to financial as opposed to technological interdependence within a firm is interesting but probably not very useful, because the structure of financial interdependence lacks the requisite stability for conventional uses of input-output. Despite the accounting conventions that provide some sort of stability to financial interdependence within a firm, many factors—outside the firm's control—are highly unstable and can play havoc with the firm's financial structure in a very short time.

Farag applies the static open input-output model to a multiproduct firm, taking the sales and inventory requirements (the final demand) as given (1968, pp. 312-320). His discussion is based on a numerical example with three production departments, using data derived from various accounting records. Unlike Richards' analysis, Farag's paper covers only the production activities of the firm, excluding financial transactions. His contribution, although competently presented, is not very original; he simply applies the static open input-output model in its conventional form to the individual firm as if it were a national economy, by simply renaming some of the variables.

Ijiri demonstrates the applicability of the static open input-output model to cost accounting in a manufacturing firm with interdependent processes (1968, pp. 49-61). He does this in terms of a hypothetical numerical example with two products and two processes. In the conclusion to his paper, Ijiri recognizes the rigidity of the assumptions of input-output in terms of the existence of by-products, joint products, and structural change in the real world, but proposes no way to deal with them. Thus, his contribution is purely pedagogical and does not break new ground in the micro application of input-output analysis.

Gambling proposes the static open input-output model as a practical tool for measuring in a consistent manner the effects on cost accounting of technological change within the firm (1968, pp. 33-38). He presents a hypothetical numerical example with two processes, each producing a single product. Before noting the theoretical shortcomings of input-output analysis caused by its simplifying assumptions, Gambling mentions the important fact that, for an actual large manufacturing company, the input-output matrix requires elements in the millions—but with many zero entries—requiring a massive computational effort.

The noteworthy aspect of his paper is its concern with the future—that is, its interest in using input-output (after predicting technological change, changes in primary factor prices and sales exogenously) to budget for future periods. Gambling realizes that static input-output analysis, which he relies on, is not very reliable in budgeting for the future. He also realizes, however, that such econometric methods as regression analysis are much less reliable because they are completely unable to accommodate drastic technological change.

Livingstone presents another version of the conventional open static input-output model in accounting terminology (1969, pp. 48-64). His discussion, based on hypothetical numerical examples, is focused on cost accounting for the individual firm. Livingstone, in this purely pedagogical paper, claims but does not demonstrate that input-output analysis can be used as a planning and control technique by a modern corporation with interdependent departments. Although he distinguishes between input-output coefficients in physical units and those in monetary units, he takes the former and product prices as given, without questioning how the corporate planners would deal with or actually induce changes in technology, factor, and product prices within the input-output framework.

Gambling and Nour, in a critical survey article, warn against the mechanical application of input-output in its macro (national) formulation to micro (firm) problems (1970, pp. 98-102). They correctly note that macro input-output coefficients in monetary units are not really technological coefficients and that they "represent an average of the actual expenditures in many production plants over an historical period of time" (1970, p. 99). They argue that the use of input-output in computation of "standard costs" of a firm should be based on the fact that they are required in *forecasting* and *controlling* future performance. They point out, however, that the input-output "technology matrix is the only feasible way of updating standard costs in very large systems of processes which are subject to continual change" (1970, p. 99).

Feltham provides an elaboration of Gambling's 1968 discussion, with greater emphasis on the activity analysis generalization of the conventional static open input-output model (1970, pp. 11-26). He presents two hypothetical numerical examples, using input-output coefficients expressed in physical units. This contribution, too, is purely pedagogical; he does not deal with whether and how a corporation in the real world could make practical use of his analysis.

Butterworth and Sigloch, in their theoretical paper sprinkled with hypothetical numerical examples, aim to develop a general multistage input-output model from which all possible alternative forms may be derived as special cases (1971, pp. 700-716). Their analysis shows clearly how the macro (national) and the micro (firm) input-output models differ in their formulation. The macro models are *simpler* because they cover the entire economic cycle so that each individual sector's monetary activity level represents both its total monetary inputs and its monetary output; that is, activities cannot be separated into inputs and outputs as they can in the micro models. The macro model is a closed system (except for international trade) but the micro model is not. This distinction raises another question. How can we show the interrelationships among different micro (open) systems within the framework of a macro (closed) system, or how can an individual firm link its input-output model, covering its own activities, to the national input-output model? Butterworth and Sigloch do not deal with this question. They note in their conclusion that it remains to apply their theoretical analysis to a particular real-life system. They merely suggest "that the levels of the input-output model may easily be identified with cost or profit centers, divisions or plants, in which case the transformations applied to the basic model may be chosen to supply the transfers and aggregations desired for planning, budgeting and control purposes" (1971, pp. 715-716).

Enrick (1971) applies the open static input-output model on the purely physical plant level. He is concerned with the "materials explosion" and aims at eliminating materials shortages or excessive stocks caused by faulty management on the plant level. He argues that a modern manufacturing plant produces many items on assembly lines, with many subassemblies and components whose interdependences are too complex to be handled systematically without input-output analysis and electronic data processing. He bases his analysis on a hypothetical (simplified) numerical example from the manufacturing of cutlery products. But he exploits only the consistency aspect of input-output, without raising any questions about technological changes and productivity adjustments.

Frank presents another application of the open static input-output model to cost accounting problems in terms of a numerical example for a purely pedagogical purpose (1974, pp. 371-376). He makes no analytical or empirical contribution.

To summarize: (a) all applications to firm and plant problems in the accounting literature have relied on the static open model and thus left out of consideration dynamic problems such as investment; (b) most have applied the macro formulation of the static open model to micro production units in a mechanical, undiscriminating way; and (c) most have ignored technological changes and productivity adjustments.

Problems in Applying Input-Output Analysis to Business Needs

Notwithstanding the rather discouraging impression the accounting literature gives for applying input-output analysis to corporate planning and control, Gols reports that several United States corporations use input-output regularly. After noting the sparseness of the literature in the existing input-output applications to corporate planning—"there is no book, no collection of essays—only a few papers" (1975, p. 21); Gols does not include the accounting literature in his references—he discusses four principal applications: forecasting, sensitivity testing, flow and structural analysis, and sorting and screening. The first two appear to have been detailed macro exercises, providing individual industries and firms with external guidelines to their planning rather than specific corporate plans. The last two types of applications seem to have been relatively few. Gols stresses their detailed macro versions, again for providing micro units with guidelines to their planning. He adds (1975, p. 23), however, that:

> There is, in addition, the possibility that flow analysis along input-output lines can be carried out for an individual enterprise in identifying flow patterns of materials between different entities within a given corporate enterprise. This is beginning to be looked at by some very large chemical corporations that have substantial intracompany product flows. It may yet turn out to be an important application for input-output analysis as more corporations move toward highly vertically integrated operations.

Among the reasons why it is taking input-output analysis so long to become a practical corporate planning tool, Gols mentions the "reluctance of input-output analysts to deal with the nitty-gritty aspects of applying input-output to business problems," (1975, p. 24) and the analyst's ignorance of corporate management. In his hypothetical numerical example—which provides far more insight than the hypothetical numerical examples of the accounting literature—Gols stresses those "nitty-gritty aspects" which involve substitutions between intermediate inputs, changes in product specifications, differential changes in input prices, changes in input specifications, changes in plant capacity utilization rates, and changes in demand for the final products as well as their interdependences. The real world, it seems, is really filled with the nitty-gritty.[16]

The Role of Time in Input-Output Analysis

Dorfman (1954, p. 123) discusses "the problem of time" in input-output analysis as follows:

> The technological relationships express the common sense fact that the quantities of raw materials used by an industry depend on its level of activity. Since the raw materials are the outputs of supplying industries, there is, at first, a direct connection between the level of activity in any industry and the levels of its supplying industries. At second glance, though, the connection is not quite so direct (even apart from the inventory problem), for, purely as a matter of technology, inputs (except for services) must be produced before they can be used. Consequently, the technological relationships (except for service inputs) connect the current level of each industry with previous levels of its supply industries and subsequent levels of its customer industries. Since the usual input-output model abstracts from the time sequence of production and exchange, it applies only to a stationary equilibrium, where time is of no consequence.[17]

By "the usual input-output model," Dorfman means the static open model. In this model, there is no lag in production; and, furthermore, the time unit is not uniquely determined by theoretical considerations. The intermediate current input requirements coefficients and the labor requirements coefficients are invariant with respect to the time unit chosen, since they are expressed as the quotient of an input rate (a flow) and an output rate (also a flow). Although in empirical input-output studies based on the static open model, the unit of time chosen is very often if not always one year, this choice is no more justified theoretically than the choice of one year in measuring the gross national product. The choice of the time unit is dictated not by theory but by the purposes of the analyst and by the resources available for collecting and processing the relevant data.

Although the intermediate current input and labor input coefficients are invariant with respect to the unit of time in the static model, the aggregate capital input coefficients are not. Aggregate capital is measured as the value of a stock (of various types of fixed capital). The aggregate capital coefficients are, therefore, stock/flow ratios, the value of capital stock divided by the capacity output per unit of time. The capital coefficients will increase by a factor of four (twelve) if the unit of time is changed from a year to a quarter (month).

The role of time is much more crucial in dynamic input-output analysis that treats capital not as a homogeneous primary factor but as a set of heterogeneous intermediate inputs used to increase sectoral production capacities over time (according to *explicit* gestation periods). As Chenery and Clark (1959, p. 17) explain:

> In the input-output framework, investment can only be explained in terms of the levels of output and capacity in each sector over time. The actual relationship between capital stock and output is subject to considerable variation, depending on whether we assume some building ahead of demand by industry, operation at some level of "economic capacity," or forced-draft operation near the maximum technological capacity. Moreover, there is a variety of ways of overcoming production bottlenecks—varying final demand, using substitute materials, working overtime, expanding existing plants, building new plants, etc. It is obvious that forecasting and planning models would usually require quite different assumptions, therefore, and that a long-run relationship may not hold for short-run variations in output.

Empirical Studies at Submacro Levels

Studies of Plant, Firm, and Industry Technologies: Pre-1960

Studies in the Structure of the American Economy, contains six empirical studies of the technological structure of the United States economy, ranging from the national level to the plant level. R.N. Grosse presents capital and inventory coefficients for 68 industries, based on data for 1939 (1953, pp. 185-242). Although he distinguishes between average and incremental capital coefficients in his theoretical discussion, Grosse's empirical estimates are mostly average coefficients based on accounting data, rather than incremental coefficients derived from engineering data or new plant construction. Grosse computes both aggregate capital coefficients and capital coefficients by industry of origin (all in dollar values). All capital coefficients measure the capital requirements per unit of estimated capacity, not actual level of output. Inventory coefficients, however, measure actual average stock levels per unit of output, not capacity levels. From his analysis of the 1939 capital structure of the United States economy, Grosse concludes that a few sectors dominate both the use and supply of fixed capital, and that the heavy users of capital are not, with few exceptions, manufacturing and mining sectors.[18]

In his brief theoretical analysis, Grosse touches on three problems of the concept of a capital coefficient: (a) because of economies or diseconomies of scale, the optimum output per unit of capital may vary even with a constant technique; (b) substitution between techniques may take place (changing the proportions between the different types of capital); and (c) the aggregation problem cannot be avoided when the coefficient is estimated for an industry rather than for a product (1953, p. 186). He claims that the first two problems are not very serious empirically but offers no opinion on the third. He then devotes his entire analysis to empirical estimation problems, the most serious of which concerns capacity estimation. His resort to different capacity estimation methods for different industries, however, raises important questions of conceptual as well as of empirical inconsistency. In response to such questions, Grosse merely cites the incompleteness and imperfections of the available data.

Clark develops a theory of private investment, called the "capital-requirements theory," as a modification of the pure accelerator principle and in a form suitable for incorporation into input-output analysis and applies it to the United States telephone industry (1953, pp. 243-294). One modification Clark makes in the accelerator principle is the allowance for technological change that requires the use of incremental as opposed to average capital coefficients for different types of capital equipment. Clark's main conclusion is that the capital-requirements theory explains the investment process of the telephone industry between 1910 and 1950 satisfactorily. He also notes two implications of his study for the development of a dynamic input-output analysis: (a) the theory works better over periods of several years than over a single year; and (b) over periods of several years especially, technological changes are drastic Therefore, dynamic analysis may have better explanatory power over periods longer than a single year (during which managerial discretion may affect investment), but for this it must keep track of and incorporate best practice techniques as they are introduced into production (1953, pp. 293-294). Clark ignores the fact, however, that over periods longer than a year changes in prices (either correlated or uncorrelated with technological changes) may affect a firm's investment. His capital-requirements theory allows technological changes but not price changes to influence corporate investment.

Chenery's well-known study derives production functions from engineering data involved in designing industrial plants, and discusses the significance of this derivation method for input-output analysis (1953, pp. 297-325). Chenery distinguishes between production functions for single processes (process functions) from production functions for an entire plant. Many processes will be independent of each other, but others will be interdependent within a given plant. (This is the basic idea underlying Butterworth and Sigloch's multistage model.) Process interdependences that can take different forms are, in general, expressed by side relations or constraints. Before and after deriving actual production (and cost) functions for natural gas transmission, Chenery notes that the engineering production function can treat explicitly changes in output dimensions and in input *qualities* whereas the neoclassical production function, estimated statistically, cannot. He concludes by suggesting that his approach can be used as an alternative way of calculating input-output coefficients and analyzing their changes due to shifts in relative prices or technological progress.

Chenery claims that the process function is a promising tool for dealing with the problem of technological change because its effects on the component functions (for transformation, energy supply, and inputs) can be separately analyzed and estimated in terms of variations in their parameters. These parameters are input and output dimensions. Chenery does not, however, concern himself with how his model can be applied to technological changes in a multiproduct plant with many interdependent processes.

Holzman's study initially poses the question of "a completely valid industrial classification for input-output purposes" (1953, pp. 326-359) and compares alternative aggregation criteria. An ideal classification must completely avoid the product and process mix problems. But such a classification scheme is either not feasible or not practical. The empirical part of Holzman's analysis is devoted to the discussion of the ninety-six-industry classification of the U.S. input-output table for 1939.

Holzman argues that the derivation of input coefficients from Chenery's engineering production functions would be superior to deriving them from aggregate transactions data by (a) alleviating the process mix and the multidimensional product problems (the latter is different from the product-mix problem; it refers to the fact that an establishment usually produces a product with different specifications which have different costs); and (b) making it possible to investigate substitutability among inputs and scale effects that create nonlinearities. Unfortunately, however, engineering functions are difficult to utilize because of "the hiatus between technical and economic data" for some but not all products; for example, the engineering production function for machining and economic data as output of an input-output metal-fabricating sector (such as electrical equipment) have a missing link that is a multidimensional frequency distribution. This distribution expresses the output dimensions of all the machined parts embodied in the finished product of the sector (1953, p. 357).

A.P. Grosse (Carter) investigates the technological structure of the United States textile industry in terms of economically meaningful production functions from engineering data (1953, pp. 360-420). She compares the engineering input coefficients with the analogous coefficients derived from "economic"—that is, observed transactions data—and explores their relative merits in the analysis of technological change. Like Holzman, she points out the difficulties in using engineering data and process functions to estimate economic relationships, referring to these difficulties as the "translation problem." Grosse distinguishes between average and best practice production functions in terms of the vintages of capital equipment. She compares the two types of functions by comparing the physical productivities (in pounds of output per unit of input during a year) for products of the same quality dimensions on the basis of data for several years between 1910 and 1946.

A.P. Grosse's analysis of changes in the average technology in terms of the *best-practice* process function does not provide a full explanation of technological change. The weights of the component functions, which depend on investment decisions, are exogenous to her analysis. Her selection of several input and output dimensions to deal with changes in qualities of inputs and outputs that are inseparable from technological change is merely suggestive rather than complete.

The sixth and last of these empirical studies is by Ferguson, who investigates the current and capital input requirements of commercial air transportation in 1939 and 1947 (1953, pp. 421-447). He restricts his estimation to fuel among material inputs and to flight crews among laborers, but considers all capital requirements. He uses a physical measure of output, available ton-miles produced per month, to compute input coefficients. He works with available capacity instead of capacity sold because the former is the chief determinant of production costs for air transportation. He converts available seat-miles to ton-miles for passenger services. The capital inputs are measured in dollars, the fuel input is measured in pounds, and the crew input is measured in both man-hours and dollars. After constructing the capital, crew, and fuel input coefficient functions, Ferguson derives the marginal cost (only of fuel cost in physical units) functions related to changes in the quantity of output, the quality of product, and the technology. Changes in product quality depend on changes in speed, altitude, and the number of landings, which are three of the determinants in the fuel consumption function. Changes in technology are specified in terms of changes in aircraft structure, design, and engine performance. Ferguson does not present a complete empirical verification of the formal input functions he constructs, but he uses data for June 1949 provided by Northeast Airlines to demonstrate that his input functions are operational. Ferguson admits that his treatment of the determinants of investment in aircraft is unsatisfactory.

These six empirical studies are impressive in their technical details and noteworthy for their common goal to bridge engineering and economic analyses. The evidence they offer for the applicability of input-output analysis to production and cost problems of individual plants and firms is, however, inconclusive. They pay inadequate attention to investment decisions that determine the direction and scope of technological change. These studies do not present input-output as a corporate planning and control tool explicitly. Instead, they aim at providing a micro foundation to the macro framework of input-output analysis (in response to criticisms that it has no such foundation). Their concern with technological change on the process, plant, and firm levels suggests, however, the possibility of employing input-output as a managerial aid in corporate decision-making. This possibility, unfortunately, has still not been demonstrated by actual experiments however, although claims for it abound. Although input-output analysis based on engineering data is feasible, it may be suboptimal as a decision-making device for a plant because its costs outweigh its benefits. It also remains doubtful whether these studies do indeed provide a firm micro foundation for the macro applications of input-output, because even on their micro levels they, by assumption, reduce the scope of substitutability of techniques, changes in product mix, and changes in qualities of inputs and outputs, and because they also fail to solve the "translation problem" (from engineering to economic data) satisfactorily.

Studies of Plant, Firm, and Industry Technologies: Post-1960

Similar empirical studies, covering petroleum refining, chemical, food and agriculture, and metals and metal-working industries, are presented in *Studies in Process Analysis*, which was published a decade after *Studies in the Structure of the American Economy*. Its editors, Manne and Markowitz, note that process analysis approaches "the analysis of industrial capability through models reflecting the structure of productive processes," and define it as "the construction and use of industry-wide, multi-industry and economy-wide models which attempt to predict production relationships on the basis of technological structure" (1963, p. 4). They stress the reliance of process analysis on engineering data and on the theories of activity analysis and linear programming. In comparing process analysis with input-output analysis, they criticize the latter for (a) ignoring alternative techniques of production (the first assumption in this chapter); (b) "the way in which 'industry output' and 'inter-industry flows' are used as basic categories of analysis," (p. 15), (that is, the product mix and the multidimensional product problems); (c) ignoring the sharing of capacities among such industries as metal-working by assuming rigid capacity constraints in feasibility analysis; and (d) ignoring joint products (the second assumption in this chapter). All except (c) of these criticisms have already been discussed. Manne and Markowitz discuss the latter as follows:

> The sharing of capacity between metalworking industries, for example, could be handled by means of "conversion coefficients" which showed the extent to which the capacity of one industry could be converted to another. Such coefficients would still fail to characterize properly the possibilities for reducing output in one set of industries to supply equipment and labor needed in another set. A more satisfactory approach is to explain the sharing of capacities in terms of the kinds of transferable resources used by these industries (1963, p. 19).

Manne and Markowitz do not, however, point out that capacity sharing through interindustry resource transfers is very much a question of time—that is, it takes time to reallocate resources—and that *static* input-output analysis which assumes fixed industrial capacities is timeless—it cannot treat time as a variable, as we have already noted. *Dynamic* input-output analysis, however, does not suffer from this deficiency.

For illustrative purposes, only one paper from *Studies in Process Analysis* need be reviewed here. Fabian presents a process-analysis model of the United States iron and steel industry during the 1950s and up to, but excluding, the introduction of the oxygen converter process (1963, pp. 237-263). The model is assigned the general task of predicting the industry's responses to exogenous variables such as the availability of certain materials and the desired product mix. It is a static model because it takes the plant and equipment as given, assuming zero net investment. Consequently, it cannot deal with the question of scale economies. It also ignores the locational aspects of production as well as the substantial differences among plants, depicting the industry as one completely integrated steel plant. Nor can his static analysis deal with investment. Although his explicit mapping of the intraindustry technological structure is noteworthy, it does not seem to provide an effective methodology for assessing the effects of technological and productivity changes that diffuse slowly and affect different plants, products, and inputs dissimilarly.

In a later study, Day follows a linear programming approach, similar to that of Fabian, to investigate the industry-wide technological changes in both the U.S. coal and the iron and steel industries (1970, pp. 99-118). Each industry's alternative production processes are represented by input-output techniques. Choice among them is based on explicit hypotheses about short-run optimizing behavior, subject to behavioral or "frictional" constraints on the replacement of one technology by another. Day assumes each input-output process to use the services of some "quasi-fixed" factors or capital good capacities. He defines "capacity" as a depreciated amount of the preceding production period's available capital stock plus current investment, assuming implicitly that the gestation period is zero. His model is then static, with a time horizon of one year. Current investment in Day's model is based on current cost differentials.

Like Fabian, Day does not specify an objective function for his model. Neither does he provide its explicit mathematical structure. Day writes that his input-output coefficients for the iron and steel industry "are based as much as possible on technical (engineering) data" (1970, p. 111). He takes Fabian's study as a starting point, adds oxygen and electrical steel production processes to his model, and pays more attention to scrap utilization and special steels. Day's reported empirical results consist of estimates of annual production levels (between 1955 and 1964) in thousands of net tons of by-product coke output, pig iron, open-hearth steel, and basic oxygen steel.

Day attributes to the industry the behavior pattern of an individual firm with respect to short-run optimization. This assumption is unrealistic (Gold,

1965, pp. 296-99). It would be theoretically defensible if *all* firms in the industry were identical as in the neoclassical theory of perfect competition. Day gives no justification for it. Furthermore, Day's model is static (with a zero gestation period), and the analysis of investment decisions by a static model is a dubious exercise, even though his treatment of investment uses annual capital changes derived from capitalized costs and payback periods—that is, dynamic considerations. What is being optimized is never made clear because the objective function is left unspecified. Day presents estimates of production levels for the industry. This is not, however, the primary purpose of his model. He states at the beginning that he is concerned with the choice of optimal technology. Does the industry as a whole or do individual firms make that choice? What about financial variables and other corporate management considerations which bear on investment decisions directly? At the end of his article, Day recognizes some of these shortcomings, but offers no clues as to how they can be remedied.

Carter, in a paper (1960, pp. 283-91) published a decade before Day's, deals with essentially the same question: how to explain changes in technology in terms of investment in new capacities? She focuses on a single industry, the tin can industry, and works with individual plant data (cross-sectional data for the 1951-1956 period compiled by the Bureau of the Census). She treats investment exogenously; that is, she does not consider what determines the rate of investment. Furthermore, she confines her analysis to *aggregate* investment on plant and equipment, without disaggregation by producing sector.

She tries to answer (statistically) the question: "To what extent is it possible to explain changes in the distribution of individual plants' (labor and electricity) input coefficients in terms of their respective equipment expenditure patterns and a common incremental or best practice production function?" (1960, p. 283).

As Carter herself points out, the success of her method in answering this question depends on the extent to which the variable inputs of an industry depend on the characteristics of its capital equipment. Her method can be viewed as a way of testing the embodied technological change hypothesis. It is not, however, a general method for analyzing and measuring technological change, which may or may not be embodied in new equipment. Furthermore, her method requires a general equilibrium formulation (which she has provided in her later work) for its increased validity. Her method is vulnerable to criticism on other grounds also: it treats the industry-wide capital coefficient as constant and uses it in the estimation of individual plants' variable input coefficients' changes; it pays no attention to changes in relative prices; it ignores all interplant differences in equipment and product mixes, in prices paid for the same input, and in other locational aspects. She is aware of most of these issues, but still thinks that a cross-sectional statistical analysis of plants is worthwhile.

Changes in input-output coefficients over time and their exogenous analysis occupy the attention of papers by Vaccara and Buzunov in *Applications of*

Input-Output Analysis. Vaccara is concerned with the coefficient changes in the United States national tables (1970, pp. 238-260). She argues that in a mature large industrial economy such as the United States, technological change in an entire industry will be slow and orderly because that change relates to new capacity which is a small proportion of the total capacity; for example, "while the newer industrial process for producing steel ingots, which involves the BOF, was perfected in 1954, as of the end of 1966, only 25 per cent of the steelmaking capacity in the United States relied upon this new process" (p. 239).[19]

After defining technological change as "change in the physical requirements for the specific goods and services used in producing a given basket of goods," (1970, p. 240) and listing it as a cause of changes in the national input-output coefficients, Vaccara lists five other causes: (a) changes in product mix; (b) non-constant returns to scale and changes in capacity utilization; (c) fictitious transfers of secondary products and competitive imports; (d) changes in the level of aggregation; and (e) random errors of measurement, changes in data sources, and statistical methods. Although Vaccara argues that "a complete analysis of coefficient change should attempt to factor out these various causes and relate them to the underlying forces of changes in consumer tastes, changes in relative prices, and technological innovation," (p. 241) she is not able to do this. That is, the identification problem of coefficient change analysis is unsolvable.

Vaccara's empirical analysis is based on the comparison of the 1947 and 1958 benchmark eighty-six order input-output intermediate input coefficient matrices, both expressed in 1958 prices. She presents several measures of the degree and direction of the coefficient changes. She finds the changes in individual coefficients erratic and unpatterned. She tries to find the reasons for coefficient changes over time by analyzing the changes in a given row (that is, from the viewpoint of the individual producing industry) instead of analyzing them along a given column (from the viewpoint of the individual consuming industry). Row analysis may be useful as a marketing-oriented approach but not as a technological change-oriented approach that requires column analysis. She realizes and partially admits this herself. But in a *national* matrix, even column analysis is of little avail in the analysis of technological change due to the basic identification problems referred to earlier.

Buzunov is concerned with the projection of the U.S.S.R. input-output coefficients along the rows of the raw materials or resource-producing industries, by either (a) econometric (regression) analysis or (b) engineering analysis (1970, pp. 322-30). He argues that (a) requires many benchmark years' data (often unavailable) and also yields average, not precise, results. On the other hand, (b) requires detailed and massive data and laborious study to yield precise results. He advocates their joint use because they are complementary. Buzunov's explicit distinction between physical and value coefficients is noteworthy because it is rarely done and, furthermore, because it makes the analysis of technological change a meaningful exercise. His analysis is restricted to the open static model, however; thus, although he examines changes between 1959 and 1970, he pays no attention to changes in capital requirements. In fact, he considers only intermediate input coefficients, completely ignoring primary inputs (that is, labor and capital). Moreover, his analysis is partial equilibrium because he focuses on a subset of industries in isolation from the rest of the economy. Therefore, his empirical conclusion that there has been a substantially more efficient use of raw materials in the Soviet Union after 1959 is not supported by convincing evidence.

Ozaki realizes that a meaningful analysis of technological change must be based on "a commodity-base approach, using cross-section and time series analyses of data for individual *production units* (plants) producing a single commodity" (1970, p. 280). He wishes to avoid the identification problem inherent in the analysis of changes in input coefficients constructed from establishment-base data. His purpose is "to determine, statistically, stable input-output technological relationships for the production process for each commodity" (p. 281). His empirical experiments are concerned with (a) testing the assumption of constant input-output coefficients with respect to plant scale for a single commodity; and (b) estimating a commodity-base input function for intermediate inputs, labor, and capital.

Ozaki measures labor in physical units (numbers of workers) but capital and material in monetary units in his cross-section analysis. He does *not* test the scale hypothesis with respect to individual commodities because he has no census input data for them. His test is with respect to *groups* of individual commodities. Therefore, his empirical results cannot apply to individual commodities. He tends to ignore this fact, however, mentioning it only briefly in a footnote.

In his time-series analysis, Ozaki measures both output and labor in physical units, but capital in monetary units. His "plant-based, factor-limitational production function" consists of three separate input functions for labor, capital, and material inputs, with capacity output as their common argument. This production function allows for variable returns to scale for each input, but excludes factor substitution for given plant capacity. Ozaki defines economies of scale in terms of the individual plant unit cost curve (for a single commodity). He measures the effects of economies of scale *independently* for each component, by estimating the three separate input functions for each commodity *group*. Ozaki's scale concept is unusual and misleading because the usual concept of scale assumes all inputs to vary by an equal proportion. His assumption that there can be no input substitution with fixed plant capacity, regardless of the length of time for which the capacity is fixed, is too restrictive, given his concern to investigate the departures from the conventional input-output assumptions. Furthermore, he fails to eliminate the product mix problem from his analysis of scale effects.

In a more recent paper (1974), Ozaki uses the theoretical framework of his earlier work (reviewed above) to explain statistically the high rate of growth of the Japanese economy, stressing technological change. As in his earlier study, he is confined to the static input-output framework, and as a result he works with the aggregate capital stock concept. Also, as before, he assumes the independence of changes in primary factor (labor and capital) coefficients from those in intermediate input coefficients. He first analyzes these changes separately and independently, and then tries to combine their results to explain the structural changes between 1955 and 1970. He concludes that "the effects of economies of scale in the use of labor and capital inputs and the substitution effects of the new for old products in the use of material inputs have played an important role in the structural change in the rapidly growing Japanese economy" (p. 7). Despite its several shortcomings, Ozaki's approach to structural change is noteworthy for its explicit concern with the scale effects.

Carter's Structural Change in the American Economy

Carter's later work on technological change is contained in her *Structural Change in the American Economy* (1970a) and a paper in *Contributions to Input-Output Analysis* (1970b, pp. 77-98). The first part of the former concentrates on the effects of structural change (used synonymously with technological change) on industrial specialization, and the second part concentrates on economic efficiency (i.e., productivity adjustments) on the macro level. The results of the first part, based on data for the 1947-1961 period, follow: (a) both capital and labor inputs per unit of fixed final demand have declined, resulting in higher productivity for both capital and labor; (b) total (direct plus indirect) intermediate inputs per unit of fixed final demand, however, have been stable, increasing slightly in some cases; and (c) sectoral input-output coefficients have been sensitive to changes in relative prices over time. Carter finds some tendency for rows of coefficients to change somewhat uniformly; for example, most input coefficients for steel and coal decline, but those for aluminum and gas rise. The second part of Carter's book, also based on 1947-1961 data, after an analysis of structural change in terms of *total* (direct plus indirect) primary factor requirements, leads to the following conclusions: (a) both capital and labor coefficients have declined in most sectors, the latter faster; (b) there is no significant relationship between increasing labor productivity and decreasing capital coefficients–Carter attributes most of the increase in labor productivity to "major qualitative changes in capital goods" that are not measurable; (c) direct primary factor saving is different from "adaptive change," which refers to reduction in primary factor requirements achieved by intermediate input substitutions based on changes in relative prices; (d) a linear programming model based on the embodied (in capital goods) technological change hypothesis is helpful in the analysis of structural change; and (e) H.A. Simon's "trigger effect" is insignificant in determining the direction and degree of technological change.

Carter presents an admirable critical discussion of the conceptual problems surrounding input-output analysis, including product mix, process mix, identification (between input substitution and technological change), classification and aggregation, and quality changes in inputs and outputs. She also criticizes the theoretical shortcomings of the static open input-output model that underlies almost all of her empirical analysis and her reliance on value (as opposed to physical) coefficients in analyzing technological change. In discussing her methodology, she asks whether direct or total (direct plus indirect) input coefficients should be favored in the analysis of structural change, and points out the index number problem involved in the choice of the common final demand vector as the basis for comparison of input productivities in different years. She is also apologetic about her capital concept and data:

> It is important to bear in mind that the capital stock coefficients available for this analysis are crude and serve only to establish orders of magnitude for the quantities under discussion. It is particularly difficult to secure capital coefficients that are comparable over time. The difficulty lies not merely in a dearth of information, but also, fundamentally, in the changing qualitative character of the capital goods composing the stock. Qualitatively different capital goods of several vintages coexist in the capital stock at any given time, and awkward fictions are required to convert the aggregate capital stock for any sector or group of sectors into common dollar units (1970a, p. 42).

In presenting her results on changes in industrial specialization, Carter (1970a, p. 42) notes that "productivity increases and changes in specialization often go hand-in-hand with the reorganization of industry for greater efficiency." In addition, she suggests that changes in the skill mix of labor, which are not dealt with in her analysis, are partly responsible for its higher productivity, measured either in physical units (man-years) or constant dollars (as total wages). Among other factors, important but not encompassed by her analysis of labor productivity adjustments, she mentions qualitative changes in the intermediate inputs and capital goods over time. Significant reductions in basic materials inputs are explained in terms of (a) cost advantage of (cheaper) new materials; (b) decreases in the size of capital goods needed to perform a given task such as miniaturization; and (c) decrease of waste through less scrap and spoilage in production that results from simplification of design and improvements in the techniques of fabrication. These two changes are not independent of changes in the qualities of the materials themselves.

Carter discusses at length the problem of qualitative changes, but concludes that it has no solution, especially for capital.

In dealing with the question of "adaptive structural change"—that is, indirect primary factor saving—and its economic significance relative to direct primary factor saving, Carter observes that "adaptive change" can involve invention as well as substitution (of one known technique for another). She admits, however, that such a distinction cannot be made in her empirical analysis. And her empirical analysis of H.A. Simon's "trigger effects"—that is, induced substitution—suggests that these are not very important, though this may be caused by her high level of sectoral aggregation. Furthermore, she notes that many sectors' structural changes are interdependent in a qualitative way, and no intertemporal comparison of conventional input-output tables can reveal that, no matter how disaggregated those tables are.

Her empirical results show the dominance of direct factors, especially labor saving, in structural change. Carter offers two reasons for the dominance of direct labor costs: (a) labor is a significant part of total cost and its price rises relative to prices of other inputs;[20] (b) intermediate input substitutions are hidden by the vertical integration of technologically interdependent activities which cannot be discovered from establishment data. Therefore, changes in direct labor (and capital) requirements are indirect evidence of how significant internal unreported changes in intermediate coefficients are. At any rate, she emphasizes that increasing labor productivity is the primary objective of the economy in both theory and practice.[21]

In a later chapter, Carter points out the lack of data on the newest or "best practice" techniques in each sector and also states her skepticism about the practical value of the embodiment hypothesis that all technological change is embodied in capital. After presenting an extensive empirical analysis using the static open input-output model with aggregate labor and capital coefficients, she criticizes the use of a static model (in which total investment is fixed) to study a dynamic problem—optimum intersectoral investment allocation—and the use of incremental coefficients (estimated from observed 1947 and 1958 coefficients with weights based on sectoral gross investment rates between 1947 and 1957) to represent new technology. She does not, however, then raise the question as to how meaningful it is to use an aggregated national input-output model in value units to study any type of technological change in its engineering or process analysis definition—a question she raises and answers negatively in other places.

A Summary of Conclusions

Carter's empirical work in structural analysis, based on the input-output and linear programming techniques, is by far the most extensive and sophisticated in its class. It also reveals most clearly (frankly) the shortcomings of the methods used, thanks to Carter's extraordinary propensity for self-criticism. Carter directly confirms the three basic conclusions which emerge from our own survey and critique.

First, the identification problems are very serious and seem to defy solution: (a) efforts to distinguish such developments as changes in factor quality and factor utilization from factor substitutions and changes in product mix are ineffective in conventional models; (b) efforts to separate changes in factor and product prices from those in factor and product qualities are ineffective in conventional models; and (c) efforts to differentiate changes in production levels from those in productivities at "full capacity" outputs are ineffective in conventional models. Efforts to deal with these problems in unconventional input-output models have also not been very encouraging so far, although there is always the possibility of future break-throughs.[22]

Second, the relevance of the analytical assumptions to the actual characteristics of the technologies and managerial practices involved is questionable in conventional models, which are almost exclusively macro-oriented. The relatively few micro-oriented models have not so far succeeded in remedying this deficiency for two basic reasons: they have failed to solve satisfactorily the identification problems mentioned in our first conclusion. In addition, they have been, on the whole, mechanical *ex post* applications of the static open model without the necessary detailed analysis of the characteristics of the micro units studied and without the development of input-output into forms suitable for serving as a serious management control and planning tool. All have failed to account for investment decisions in a realistic way and thus cannot focus on the critical interdependence between investment and technological change.

Third, the applicability of empirical results to macro level questions is clearly greater than to management decision-making. *Ex post* analysis of national structural changes by itself may be useful in an economy lacking a national planning mechanism, but *ex post* analysis of corporate structural changes, involving plant level technological changes, by itself is clearly inadequate. Such technological changes do not just happen. They are planned for and often have a critical importance for the corporate decision-makers. Thus, they must be amenable to *ex ante* analysis by input-output, if input-output is to make a significant contribution to management decision-making. This has not yet happened.

Notes

1. Concerning the compatibility of input-output and linear programming analyses, Chenery and Clark (1959, p. 334) noted:

> The essential point, however, is that there is no inherent contradiction between the two approaches, and they can be combined whenever it seems both desirable and practicable. A given set of parameters in an input-output model represents one of many possible solutions to a more general linear programming model. The choice of activities can be either predetermined or based on a criterion function, or these alternatives can be used for different sectors in the same model. Such combined models seem promising for future applications.

Indeed, since 1959, such combined models have been formulated and empirically implemented as applications in program formulation as well as in structural analysis. For examples of program formulation applications, see Murakami, Tokoyama and Tsuki (1970, pp. 24-47); Bruno, Dougherty, and Frankel (1970, pp. 48-69); and Manne and Weisskopf (1970, pp. 70-102).

2. For an earlier and theoretical exposition, see Leontief (1951, pp. 53-90).

3. Robert Dorfman (1954, pp. 128-130) discusses this diversity in theoretical and empirical terms.

4. Leontief does not confine his productivity analysis to the static open model and touches upon dynamic issues as well. But his coverage of dynamic issues does not constitute the formulation of a full-fledged dynamic model (1951, p. 61).

5. Let l = the vector of direct labor requirements per unit of gross output of n sectors and $\mathbf{A}$ = the square matrix of direct flow intermediate requirements per unit of gross output of n sectors. Then $l'(1 - \mathbf{A})^{-1}$ = the vector of total (direct plus indirect) labor requirements per unit of final demand of n sectors.

6. For a criticism of the neoclassical production theory in this respect, see Chapter 1.

7. For an extended and systematic discussion of these and related issues, see Chapter 4.

8. In the static closed model, labor becomes a produced or intermediate input and capital is the only primary factor. The static closed model, based on the Malthusian view of population and production, gave way to the static open model shortly after the advent of the input-output analysis.

9. For a detailed discussion of these problems and of possible means of dealing with them, see Gold (1955, pp. 18-30). For a parallel discussion of purposive measures of labor inputs, see Gold (1955, pp. 15-18).

10. For a discussion of further problems, see Gold (1965, pp. 299-303).

11. Other conceptual problems in this area are discussed in Gold (1965, pp. 303-305).

12. The implications of different aggregation levels for the development of distinctive objective functions and analytical frameworks is discussed in Gold (1965, pp. 295-99).

13. For an interesting alternative approach to the conceptual bases for industry and product, see Brunner (1976).

14. For a broader discussion of differences between the analytical interests of industrial decision-makers and academic (as well as many government) economists, see Gold (1971, pp. 1-18).

15. But mobilization planning has demonstrated a need for such tools.

16. For a discussion of the theoretical and empirical difficulties involved in

the application of input-output analysis in capital goods industries by individual firms, see Machinery and Allied Products Institute (1970).

17. The question of consistency in the time coverage of inputs and outputs also raises questions concerning other aspects of consistency, such as those discussed in Gold (1965, pp. 290-91).

18. "In 1939, practically all manufacturing and mining activity was carried on with some 20 per cent of the fixed capital stock" (Grosse, 1953, p. 225): "about 75 per cent of the capital stock of the economy was produced by only four industries (construction, transportation equipment, not elsewhere classified, electrical equipment, not elsewhere classified, and machine tools)." Also see p. 242.

19. For a broader perspective on empirical diffusion rates for an array of innovations, see Gold, Peirce, and Rosseger (1963, pp. 219-25).

20. For evidence of the quite different perspectives on labor costs at the level of individual industries, see Gold (1971, p. 137).

21. The inapplicability of this approach in a wide array of industries where labor cannot increase productive capacity or operating efficiency substantially is discussed in Gold (1955, pp. 188-96).

22. However, these problems present serious challenges in economic measurement in general, instead of being unique to input-output analysis. And they are sharply intensified by technological innovations, as is discussed in Gold (1973, pp. 181-90).

References

Brozen, Yale, "Comments on 'Effects of Technological Change in a Linear Model' by H.A. Simon," in *Activity Analysis of Production and Allocation*, ed. T.C. Koopmans (New York: Wiley, 1951).

Brunner, E., and P.W.S. Andrews, "Industrial Analysis Revisited," *Studies in Pricing* (London: MacMillan, 1976) chap. 3.

Bruno, M., C. Dougherty, and M. Frankel, "Dynamic Input-Output Trade and Development," in *Proceedings of the Fourth International Conference on Input-Output Techniques–1968*, ed. Anne P. Carter (New York: Elsevier, 1970).

Butterworth, John E., and Berndt A. Sigloch, "A Generalized Multi-Stage Input-Output Model and Some Derived Equivalent System," *Accounting Review* (October 1971).

Buzonov, R.A., "Technical Economic Projection of Coefficients of Direct Material Inputs," in *Proceedings of the Fourth International Conference on Input-Output Techniques–1968*, ed. Anne P. Carter (New York: Elsevier, 1970).

Carter, Anne P., "Investment, Capacity Utilization, and Changes in Input Structure in the Tin Can Industry," *Review of Economic Studies* 42 (August 1960).

_______, *Structural Change in the American Economy* (Cambridge, Mass.: Harvard University Press, 1970a).

_______, "A Linear Programming System Analyzing Embodied Technological Change," in *Proceedings of the Fourth International Conference on Input-Output Techniques–1968*, ed. Anne P. Carter (New York: Elsevier, 1970b).

_______, ed., *Proceedings of the Fourth International Conference on Input-Output Techniques–1968, Contributions to Input-Output Analysis*, Vol. 1; *Applications of Input-Output Analysis*, Vol. 2 (New York: American Elsevier Publishing Co., 1970).

Chenery, Hollis B., "Process and Production Functions from Engineering Data," in *Studies in the Structure of the American Economy–Theoretical and Empirical Explorations in Input-Output Analysis* ed. Wassily Leontief (New York: Oxford University Press, 1953).

_______, and P.G. Clark, *Interindustry Economics* (New York: Wiley, 1959).

Christ, Carl F., "A Review of Input-Output Analysis," in National Bureau of Economics Research, *Input-Output Analysis* (1955).

Clark, Paul G., "The Telephone Industry–A Study in Private Investment," in *Studies in the Structure of the American Economy–Theoretical and Empirical Explorations in Input-Output Analysis*, ed. Wassily Leontief (New York: Oxford University Press, 1953).

Day, Richard H., "Recursive Programming Models of Industrial Development and Technological Changes," in *Proceedings of the Fourth International Conference on Input-Output Techniques–1968,* ed. Anne P. Carter (New York: Elsevier, 1970).

Dorfman, Robert, "The Nature and Significance of Input-Output," *Review of Economics and Statistics* 36 (May 1954). Quoted with permission.

Enrich, Norbert L., "Materials Management by Input-Output Analysis," *Manufacturing Engineering and Management* (October 1971).

Evans, W.D., "Indexes of Labor Productivity as a Partial Measure of Technological Change," in Netherlands Economic Institute, *Input-Output Relations* (1953).

_______, and M. Hoffenberg, "The Nature and Uses of Interindustry Relations Data and Methods," in *National Bureau of Economic Research, Input-Output Analysis* (1955).

Fabian, Tibor, "Process Analysis of the U.S. Iron and Steel Industry," in *Studies in Process Analysis*, ed. Alan S. Manne and Harry M. Markowitz (New York: Wiley, 1963).

Farag, Shawski M., "A Planning Model for the Divisionalized Enterprise," *Accounting Review* (April 1968).

Feltham, Gerald A., "Some Quantitative Approaches to Planning for Multiproduct Production System," *Accounting Review* (January 1970).

Ferguson, Allen R., "Commercial Air Transportation in the United States," in *Studies in the Structure of the American Economy* (1953).

Fisher, Walter D., "Criteria for Aggregation in Input-Output Analysis," *Review of Economics and Statistics* 40 (August 1958). Quoted with permission.

Frank, Werner G., "Solving Financial Planning Problems Using Input-Output Models," *Accounting Review* (April 1974).

Gambling, Trevor E., "A Technological Model for Use in Input-Output Analysis and Cost Accounting," *Management Accounting* (December 1968).

_______, and Ahmed Nour, "A Note on Input-Output Analysis: Its Use in Macro-Economics and Micro-Economics," *Accounting Review* (January 1970).

Gigantes, T., "The Representation of Technology in Input-Output Systems," in *Proceedings of the Fourth International Conference on Input-Output Techniques–1968*, ed. Anne P. Carter (New York: Elsevier, 1970).

Gold, Bela, *Foundations of Productivity Analysis* (Pittsburgh: University of Pittsburgh Press, 1955).

_______, "Productivity Analysis and System Coherence," *Operational Research Quarterly* (September 1965).

_______, *Explorations in Managerial Economics: Productivity, Costs, Technology and Growth* (New York: Basic Books, 1971).

_______, "The Impact of Technological Innovations: Concepts and Measures," *Omega* (April 1973).

_______, W.S. Peirce, and G. Rosseger, "Diffusion of Major Technological Innovations in U.S. Iron and Steel Manufacturing," *Journal of Industrial Economics* (July 1970).

Gols, A. George, "The Use of Input-Output in Industrial Planning," *Business Economics* (May 1975). Quoted with permission.

Grosse, Robert T., "The Structure of Capital," in *Studies in the Structure of the American Economy* (1953).

Grosse, Anne P., "The Technological Structure of the Cotton Textile Industry," in *Studies in the Structure of the American Economy* (1953).

Holzman, Mathilda, "Problems of Classification and Aggregation," in *Studies in the Structure of the American Economy* (1953).

Ijiri, Yuji, "An Application of Input-Output Analysis to Some Problems in Cost Accounting," *Management Accounting* (April 1968).

Kamien, M.I., and N.L. Schwartz, "Market Structure and Innovation: A Survey," *Journal of Economic Literature* 13 (March 1975).

Kennedy, C., and A.P. Thirlwall, "Technical Progress: A Survey," *Economic Journal* 82 (March 1972).

Klein, Lawrence R., "On the Interpretation of Professor Leontief's System," *Review of Economic Studies* 20, no. 2 (1952-53).

Koopmans, T.C., ed., *Activity Analysis of Production and Allocation* (New York: Wiley, 1951).

Leontief, Wassily, *The Structure of American Economy, 1913-1939—An Empirical Application of Equilibrium Analysis* (New York: Oxford University Press, 1951).

_______, ed., *Studies in the Structure of the American Economy—Theoretical and Empirical Explorations in Input-Output Analysis* (New York: Oxford University Press, 1953).

_______, "Dynamic Analysis," and "Structural Change," in *Studies in the Structure of the American Economy* (1953).

_______, "Some Basic Problems of Empirical Input-Output Analysis," in *Input-Output Analysis* (1955).

_______, "An Alternative to Aggregation in Input-Output Analysis and National Accounts," *Review of Economics and Statistics* 49 (August 1967).

_______, "The Dynamic Inverse," in *Proceedings of the Fourth International Conference* (1970).

_______, "Environmental Repercussions and the Economic Structure: An Input-Output Approach," *Review of Economics and Statistics* 52 (August 1970).

Livingstone, John L., "Input-Output Analysis for Cost Accounting, Planning and Control," *Accounting Review* (January 1969).

Machinery and Allied Products Institute, *Input-Output Analysis and Economic Forecasting* memorandum (March 1970).

Manne, Alan S., and Harry M. Markowitz, eds., *Studies in Process Analysis* (New York: Wiley, 1963).

Manne, Alan S., and T.E. Weisskopf, "A Dynamic Multisectoral Model for India," in *Proceedings of the Fourth International Conference* (1970).

Murakami, Y., K. Tokoyama, and J. Tsukui, "Efficient Paths of Accumulation and the Turnpike of the Japanese Economy," in *Proceedings of the Fourth International Conference* (1970).

National Bureau of Economic Research, *Input-Output Analysis: An Appraisal—Studies in Income and Wealth*, vol. 18 (Princeton, N.J.: Princeton University Press, 1955).

The Netherlands Economic Institute, *Input-Output Relations* (Leiden: H.E. Stenfert Kroese N.V., 1953).

Ozaki, Iwao, "Economies of Scale and Input-Output Coefficients," in *Proceedings of the Fourth International Conference* (1970).

_______, "The Effects of Technological Changes on the Economic Growth of Japan, 1955-1970," mimeographed paper (1974).

Richards, Allen B., "Input-Output Accounting for Business," *Accounting Review* (July 1960).

Vaccara, Beatrice N., "Changes Over Time in Input-Output Coefficients for the United States," in *Proceedings of the Fourth International Conference* (1970).

4 Diffusion of Technology in Industry

Gerhard Rosegger

This chapter reviews some of the conceptual and methodological issues raised by a review of published research on diffusion; it does so specifically from the standpoint of their relevance to understanding the diffusion of technological (process) innovations in industry. As a starting point, we shall focus on the extent to which the research tradition of the anthropologists and sociologists, the pioneers in the study of diffusion, is useful in the framework of economic investigations. Our concern will be primarily with the direct applicability of this tradition's paradigms to the actual diffusion of specific innovations, rather than with their relevance to some "general" micro theory of technological change.

Experience in this field has amply demonstrated that increments in generality have always been obtained at the expense of applicability to real choices confronted by decision-makers.

The Measurement of Diffusion

The concept of diffusion is a familiar one in the natural sciences, and it is from there that the social sciences have borrowed the term. As Rapoport (1956) pointed out:

> The use of the word "diffusion" in the social sciences indicates the awareness of some similarity between the spread of, say, a technological artifact and the spread of a solute through a solvent. Of course it may be argued that the use of a metaphor does not establish the reality of the connection between situations compared and, in fact, may be seriously misleading. . . .

Be that as it may, several distinctive research traditions on diffusion were developed by social scientists. In anthropology, the main concern has been with the transmission of the features of one culture to another and with the stimuli and retardants operating on this process.[1] Sociologists have dealt more broadly with the spread of ideas, information, or rumors within some defined population.[2] Studies in this tradition have examined diffusion not only as a temporal, but also as a spatial phenomenon, with the latter subject occupying social geographers as well.[3]

A useful characterization of this work's general thrust is provided by Katz et al. (1963), according to whom the "standard diffusion model" involves (a) acceptance, (b) over time, (c) of some specific item—an idea or practice, (d) by individuals, groups, or other adopting units, linked to (e) specific channels of communication, (f) to a social system, and (g) to a given system of values or culture.

Understandably enough, this "accounting scheme" for the classification of anthropological and sociological literature, as the above authors call it, does not mention the variables an economist would look for in studying diffusion.[4] Indeed, the orientation of diffusion research in anthropology and sociology is primarily toward its informational, communicative, and interpersonal aspects, as well as toward social and cultural factors in resistance to change. This orientation manifests itself most strongly in the pervasiveness of *contagion models* for measuring diffusion. Such models, based on the paradigm of "spread by contact," have been examined in detail for their implications in the social sciences by Rapoport and Rebhuhn (1952), Landau and Rapoport (1953), and Rapoport (1956). In their most general form

$$\frac{dx}{dt} = ax(1 - x) \tag{4.1}$$

these models give rise to the familiar logistic set of diffusion curves. Here x is the proportion of all individuals (or units) already in contact with a piece of information, and the parameter a indicates what Rapoport calls "the potency of spread."

When the relevance of this research tradition to economic research into diffusion is examined, it is clear that the economist's interest in spread phenomena has three main and quite distinct foci. The first deals with the speed with which given innovations spread from enterprise to enterprise, in an effort to relate the rate of diffusion to such factors as the structure of an industry, the characteristics of individual firms (large vs. small, concentrated vs. diversified, etc.), and the nature of the innovation.[5]

A second focus of economic research is the growth of *new* products and new industries rooted in major innovations. Here the rise in the absolute value of outputs, in their percentage shares of GNP, or of some other aggregate measure, are taken to be indicators of an innovation's spread.[6]

A third focus is the extent to which innovations, especially process innovations, affect established industries by replacing or displacing predecessor techniques or by creating incremental productive capabilities. The orientation here is on output effects, in line with the economist's traditional concern with results rather than with motivation (Alchian, 1950). The pioneering example for this type of research is Griliches (1957).

It would seem clear that the conceptual framework of the contagion model fits, at best, the first of these research foci. Even here, however, there remains a methodological difficulty: the structure of the model is such that "spread by contagion" takes places within some defined entity, "the population of potential adopters." But an industry does not consist of a static (or predictably changing) population of firms. Indeed, technological innovations are recognized as one of the chief means of entry for new firms; the failure to adopt an innovation may lead to the forced exit of firms, at least with respect to the relevant market. Moreover, not all firms in an industry, as usually defined, make the same products or use identical processes, but they will consider shifts in either or both under certain conditions.

Thus, who is a potential adopter of a new technique (and to what extent of his total capacity) becomes itself an empirical question and not a matter of prior definition. The problem is compounded by what Evenson (1974) has called the "technology specificity" of production conditions in individual firms—that is, the extent to which existing technical arrangements determine whether an innovation is of interest or not.[7]

A further question must be raised about the economic meaning of the parameter indicating the "potency of spread." In the industrial setting, the decision by a firm to introduce a certain innovation does not, by itself, tell anything about economically significant effects. What fraction of total output represents a commitment to the innovation? How fully were firms utilizing existing capacities before and after the introduction? When all firms have introduced the innovation, does this mean that older techniques and products have been displaced entirely? These questions are clearly important in an assessment of an innovation's economic impact.

In a classic study in rural sociology, for example, Ryan and Gross (1943) examined the time path of the *percentage* of farmers in two communities "accepting" hybrid seed corn. No attempt was made to introduce any economic factors into the explanation of spread, and effects on total output were not considered. But as Nelson et al. (1967, p. 102) pointed out later on, none of the early adopters counted in the Ryan and Gross study planted more than 15 percent of his total acreage in hybrid corn. The percentage of adopters thus revealed nothing about the innovation's actual impact on the quantity and quality of total output.[8] In this important sense, then, the *economic* information contained in diffusion models that center on adopting units is limited.[9]

Measures and models that deal with shares of aggregate spending absorbed by an innovation clearly are not very useful for the study of established, as differentiated from new, industries. However, as Gold (1964) has pointed out, expectations about the diffusion patterns of technologies based on new industries frequently tend to overlap conceptually with expectations concerning the growth pattern of such industries.[10] But despite the prevailing tendency of popular attention to center almost exclusively on the spectacular new products and technologies, a tendency often encouraged by business and government, the fact remains that the bulk of economic progress via technological changes is rooted in major and minor advances achieved by well-established industries supplying well-known products.[11] Therefore, economic diffusion research must recognize the additive, displacing or replacing roles of innovations in these industries, if an undistorted picture of their impact is to emerge. And output-oriented measures of diffusion seems to provide the best bases for such assessments.

By output-oriented measures we mean those which either directly or indirectly (for example, by observations of "old" and "new" capacities) constitute evidence on the extent to which industrial innovations are affecting the performance of a given sector of the economy, and on the speed with which these effects permeate the sector. If diffusion research is to aid in technology assessment and in managerial planning, measures of this type are necessary to answer such questions as the following: are certain types of industry, or types of innovation, characterized by similar patterns of diffusion? Do early diffusion patterns give some indications of the eventual importance of an innovation? Does the overall growth rate or contraction of an industry's output affect rates of diffusion?

Recent empirical research[12] shows an increasing recognition of the analytical and predictive value of economic diffusion measures, with a corresponding decline in reliance on models whose conceptual framework is more appropriate to sociological and anthropological inquiry. In the end, of course, it is not doctrinaire adherence to the paradigms of any one science, but an evaluation of the specific conceptual implications of the several research traditions, that will determine the fruitfulness of empirical investigations. Thus, for example, studies of the role of patents in diffusion,[13] of the impact of differences in "managerial styles" on adoption decisions, and of the role of individual mobility in interfirm or intrafirm diffusion,[14] necessarily base their methodologies on observations of spread among units rather than on spread in terms of economic magnitudes.

Factors Influencing Diffusion Rates

In attempting to assess the several research traditions' relevance to explanations of differences in industrial diffusion rates, we shall restrict ourselves to an examination of the role of *information flows* and of the concept of *resistance.* Our concern here is with the degree to which the spread of innovations in industry is, or is not, analogous to the setting postulated in the "standard diffusion model" of anthropology and sociology, as characterized earlier.

The Role of Information

In the anthropological and sociological tradition, information is perceived as a series of messages, flowing from "origin nodes" through structured or unstructured channels of communication. In economics, information constitutes the output of certain types of productive activity. Some important conceptual distinctions derive from these different ways of looking at the same phenomenon.

At the most obvious level, in many noneconomic contexts the transmission of information constitutes the diffusion process itself, as in the spread of a rumor or of a piece of news. In most other situations, however, the receipt of information is only the initial step in a series of events that, together, constitute the observable diffusion process. The diffusion of innovations in industry clearly belongs to this latter category. Carlson (1967) put the matter very succinctly when he argued that "the mere verbal transmission of technical information which is not utilized is not diffusion of technology."

Therefore, it is probably easy to overrate the importance of such factors as the movement of people or the role of publication in its various forms as an explanation of technology diffusion. The attribution of dominance to these factors generally rests on *ex post* evaluations of successful transfers.[15] Our research on the diffusion of innovations in one industry (Gold, Peirce, and Rosegger, 1970) suggests that typically a huge amount of "information," such as technical papers or exchanges of visits by experts, so far precedes an actual rise in diffusion rates as to put into serious question the value of informational spread, as such, as an explanatory variable.[16]

Economic theory provides little guidance on the question because, as Stigler (1961) points out, the prevailing assumption of micro theory is that decision-makers fully "know" the characteristics of all techniques available at any given moment in time.[17] However, at least the following considerations seem to militate against any such facile assumption in the industrial setting, even if the problem of uncertainty is ignored in the conventional sense:

1. Though scientific knowledge as well as large amounts of technical information are in the nature of public goods, these are rarely sufficient as a basis for managerial decisions. This component of the total information system, by its nature, deals either with abstractions or with the current state of applied knowledge *somewhere else*, and therefore gives no more than clues to decision-makers about the changes necessary in their own production setup.[18]

2. In contrast to most of the information flowing through various channels of communication in other social settings, crucial information for the firm is not "free." The acquisition of knowledge necessary to a decision to adopt an innovation involves several types of costs, among them that of purchasing proprietary information (such as that contained in patents), of acquiring the appropriate capital equipment, and of learning. In an even more important sense, "sufficient" knowledge is not acquired once and for all, and a decision taken; rather, the accumulation of information proceeds in a series of decision cycles and reevaluations, many of which occur after the basic "adoption decision."[19]

3. But while acquiring information in the industrial setting for innovation

may be costly, these costs are often mitigated by the fact that third parties have an economic interest in the diffusion of innovations. Equipment manufacturers, for example, can be counted on to provide information services to firms both before and after the decision to adopt a certain technique; materials suppliers as well as customers may have a direct stake in a new process or product.

4. Because of these outside interests, some of them conflicting, the information supplied at the early stages of an innovation's commercial life may be anything but unequivocal. This is so because of the variegated experiences of early adopters, and because industrial process innovations, especially, tend to be embodied in a variety of competing equipment types. Each of these ostensibly solves the same technical problem, but during diffusion only a few survive as recognizably superior. Therefore, even a major innovation does not generally emerge in the form of a single "best technique" of the kind postulated, for example, by Salter (1960) in his model of diffusion. A phase of slow rates of spread may characterize such innovations before greater technical clarity leads to a takeoff for prolonged periods.[20]

We are left, then, with the conclusion that in the industrial setting the acquisition of information is itself such an integral part of the firm's economic activity that it would be misleading to assign the quantity and quality of information to the set of exogenous variables impinging on decisions.

The Meaning of Resistance

In the anthropological and sociological tradition, resistance to change or to the utilization of new information is rooted in cultural factors in the social structure of the population concerned and in the behavioral characteristics of individual adopters and nonadopters. Therefore, frequently used explanatory variables include such measures as the level of education of decision-makers, their "progressiveness" (however defined), the role of "social dynamics," of community leaders, opinion makers, and so on. Such variables, important as they may be in the general social framework, appear to have more readily quantifiable analogues in the diffusion of new processes and products from the economic perspective.[21]

As Kennedy and Thirlwall (1972) argue, it is necessary in economic explanations of diffusion to differentiate among the *opportunity*, the *capacity*, and the *pressure* to innovate. Opportunities clearly have to be culled from the vast amounts of information pouring in on decision-makers. According to Gold's empirically supported argument (1969), the capacity to innovate does not depend so much on "innovative" entrepreneurs (successful innovations always resulting from, *ex post*, the "venturesome spirit" of management) as on the availability of the necessary financial and physical resources and on the technical setting of the firm.[22] Finally, pressures manifest themselves in the usual form of potential cost penalties or competitive disadvantages.

Economic factors in "resistance," therefore, are likely to include a commitment to investment in the firm's present equipment and production setup; the commitment to existing sources of materials and other variable inputs; and the commitment to an integrated production system in whose total context the potential benefits of an innovation (at any one stage of production) must be evaluated.

Mansfield (1968) postulates the following, more general, determinants of the rate of diffusion: (a) the extent of the economic advantage of the innovation over older methods or products; (b) the extent of uncertainty associated with the innovation when it first appears; (c) the extent of commitment required to try out the innovation; and (d) the rate of reduction of the initial uncertainty regarding the innovation's performance. It seems clear that his factors (b) and (d) deal primarily with the role of information, which we discussed above; factors (a) and (c) are more directly related to the present context. It would also appear that the empirical validation of (b) and (d) faces serious methodological problems.

Separate attention must be given to those innovations that tend to affect the performance of the total production system, such as new methods and techniques of managerial coordination and control. Whenever such innovations are superimposed on a complex network of existing production stages, intermediate product flows, work-in-process inventories, organizational arrangements, and managerial responsibilities, the opportunity cost of failure involves not just the cost of investment or of problems in one component, but the consequences of a disruption of the entire system or of a serious loss of control.[23] At this point behavioral explanations of resistance within the firm may assume primary importance, because the success or failure of these innovations, more than any others, depends to a large extent on the attitudes and cooperation of the people involved in making them work.

Finally, if diffusion research is to provide some guidance to policy assessment and planning, it must take into account whatever institutional factors might influence the rate at which innovations are adopted. What these factors are is once again an empirical question, rather than one of some constraints postulated *ex ante*. Nevertheless, a few general observations may be appropriate. One of these concerns the role of labor unions in the technological change process. While this role has received considerable attention as a separate subject (for example, see Weinstein, 1965), surprisingly few of these findings seems to have spilled over into research on diffusion and resistance. Similarly, examinations of the influence of governmental authority on the spread of innovations appear largely restricted to rather narrowly defined subjects, such as the patent system, the differential role of technological change in the regulated industries (for example, Capron, 1971) and to the impact of environmental and consumer protection legislation. Finally, scant recognition has been given to the part which the numerous official and quasi-official specification-writing agencies play in determining the speed with which new products or the products of new processes are accepted.

Even from this limited list of economic and institutional considerations, one may conclude that an examination of the reasons for an otherwise frequently inexplicable "resistance" of firms to innovations can be conducted within the framework of the economist's concepts and variables. To be useful for strategic purposes, such research also requires an understanding of both the characteristics of the innovation in question and of the candidate firms' technological, financial, and other operating characteristics.[24] Otherwise, there is always the danger that quick resort to extra-economic explanations will produce as many misleading conclusions as could mere statistics-grinding in "general" models of diffusion. The ultimate impact of an innovation in the economic setting must, in any event, be assessed in terms of the decision-makers' frame of reference, which deals in costs, risks, and potential payoffs.

On the Diffusion of Major Consumer Product Innovations

There seem to be three major differences among the factors affecting the diffusion of major technological innovations among industrial users and those affecting the diffusion of consumer products embodying major technological innovations.[25] First, the primary criteria for industrial decision appear to be more generally susceptible to objective definition, although less objective criteria may also play a role when the evaluations of the former are not decisive. Second, the evaluation processes applied in industry tend to be more intensive and to make greater use of specialized expertise. And third, such decisions in industry are likely to have more serious consequences, both in terms of their effects on the economic well-being of the firm and in terms of the magnitude of the resource commitment involved. Indeed, it is the latter characteristic that accentuates the other two sources of differentiation.

Of course, consumer products cover a broad range in respect to each of these three characteristics. Thus, the more serious the consequences of adoption decisions to the consumers, the greater the likelihood that they will seek, and even pay for, expert guidance capable of making sophisticated judgments in respect to relevant objective criteria. This is most obvious in the case of critical health problems when decisions about therapeutic procedures and drugs are turned over to physicians. When decisions involve substantial but less serious consequences, considerable reliance may be placed on intermediaries, including companies and even individual salesmen who have engendered trust over extended periods by fulfilling their promises of product quality and reliability. In addition, friends whose advice and judgments have been supported by past experience provide another means of supplementing the consumer's own knowledge and experience. There remains, however, a vast area of consumer responses to technologically innovative products in which expert judgments in respect to objective criteria of economic and service values play an essentially minor role relative to the bundle of subjective motivations and preferences encompassed by the term "tastes."

In short, such economic considerations as income levels and product prices provide some constraints within which noneconomic influences have free play in shaping the diffusion of technologically innovative consumer products. It also seems plausible that the likelihood of satisfactory technical performance provides some additional constraints, although the bases for making such estimates in advance of purchasing a new product obviously include some noneconomic as well as nonexperiential considerations. To cover such considerations, the influence of which would seem to be dominant within the broad range of choices available in most product sectors despite the above economic and technical constraints, requires supplementing the industrial framework of diffusion analysis. Specifically, it is necessary to add back some of the very basic concepts developed by social scientists concerned with diffusion processes in various sectors of human behavior that were found to have limited relevance in explaining major innovational decisions in industry. The resulting need to consider such factors as information channels, peer group pressures, status symbols, and similar behavioral influences has been amply demonstrated by the pragmatic explanations and applications by specialists in market research, advertising, and sales.

Having emphasized these basic differences in explanations of industrial and consumer diffusion processes, we can return to an emphasis on their conceptual unity: both seek to account for the *physical* spread of some innovation among potential adopters. It is necessary to emphasize this seemingly obvious point to differentiate studies of the diffusion of consumer products from those dealing with the economic growth rates of these same products. Although the latter are frequently labeled as "diffusion research," they involve such measures as the change in consumer spending for an innovation (either in dollar amounts or in relative terms), thereby confounding price and quantity effects over time. Thus, especially in situations where the enlargement of a product market is accompanied by (or even occasioned by) continuous price reductions, economic growth rates are likely to understate actual rates of diffusion, even if the demand for the product should turn out to be price-elastic.

Summary and Conclusions

We have considered some conceptual and methodological questions concerning research into the diffusion of innovations in industry. We are led to the conclusion that, in this setting, the rich texture of technological change may yield valuable understanding for planners and decision-makers, even if research initially restricts itself to those aspects of the diffusion process that can be observed and analyzed in the context of the economist's traditional variables.

And we judge that, as an area for serious empirical research, this represents one of the underdeveloped sectors of economics, with elegant hypotheses far outnumbering careful analytical tests.

Notes

1. For a comprehensive survey that also provides a historical guide to anthropological literature, see Barnett (1953).

2. The classic example of sociological approaches to diffusion phenomena is Rogers (1962).

3. See, for example, Hägerstrand (1967).

4. Neither are any distinctive economic factors referred to in Krober's article on "Diffusionism" in the *Encyclopedia of the Social Sciences* or in Hägerstrand's essay on "The Diffusion of Innovations" in the *International Encyclopedia of the Social Sciences.* Barnett (1953) introduces the concept of "novelty costs" and concludes that "it is relatively easy to get people to try something if it entails no loss to them." But his elaboration shows that he is not thinking of the economist's cost concept: "Cost is a relative judgment. It is a function of an individual's economy of preferences. . . . The man who can afford a yacht may still find the price of a lawn mower excessive . . . " (p. 361).

5. For the most widely recognized work along these lines, see Mansfield (1968).

6. A comprehensive study of twenty product innovations, along these lines, was undertaken by Lynn (1966).

7. A detailed model, describing the decision process at the level of the firm, was developed by Gold, Peirce, and Rosegger (1970).

8. In the industrial setting, Mansfield's study (1963) is an interesting effort to explain intrafirm rates of diffusion.

9. Peirce (1974) illustrates that such models may actually be misleading: he found that, in terms of output, an upward trend in diffusion may be interrupted by prolonged periods of decline, even though no firms "dropped" a new technique.

10. For an empirical application in this vein, see Chow (1967).

11. Efforts to attribute special economic significance to distinctions between "high" and "low" technologies have further obscured this fact.

12. See, for example, Ray (1969); Gold, Peirce, and Rosegger (1970); and Nabseth and Ray (1974).

13. See, for example, Taylor and Silberston (1970) and Ifo-Institut (1974).

14. Several studies along these lines can be found in Gruber and Marquis (1969).

15. It is perhaps significant that much of the research on the importance of "human factors" and "information transfer" has been carried out in the context of government-financed projects, where the purely economic constraints on diffusion are less severe than in private sector undertakings.

16. It is an interesting question whether the accumulation of information proceeds to a point where decision-makers perceive definite pressure "to do something." The existence of technical monitoring functions in many firms suggests such a process, but the empirical evidence is limited. For some hints in this direction, see Hiestand (1974).

17. Similar assumptions in trade and growth theory have been criticized, among others, by Nelson (1968).

18. See, for example, the discussion of the transitory role of highly specific technical literature by Knoerr (1963).

19. For an experienced practitioner's description of the process, see Blickwede (1972). In a study of twenty projects involving over 800 identifiable technical events, Carlson (1967) found that approximately one-third of these events occurred *after* the projects had been initiated on a "commercial" basis.

20. The utilization of oxygen in steelmaking is an excellent illustration of this type of spread. Considerable time passed before one of four basic, and quite different, approaches established itself as superior.

21. This methodological issue is discussed fully in the well-known controversy between Rogers and Havens (1961, p. 22) and Griliches (1962). As Keynes's famous reference to "animal spirits" (1936, p. 161) shows, however, economists do not hesitate to resort to psychological or cultural explanations in the face of otherwise puzzling behavior. A rigorous treatment of managerial attitudes toward innovations, in business mythology and empirical fact, can be found in Gold (1969). The economic literature bearing on the subject is reviewed in Kennedy and Thirwall (1972). For an economic historian's balanced evaluation, see Rosenberg (1972).

22. As soon as we move out of the artificial construct of "the firm" as a unified system, the importance of rates of capacity utilization in *specific components* of the system, for example, assumes its proper significance in the firm's attitude toward innovations affecting those components. This fact is fully recognized in the normative models of activity analysis (such as linear programming). It is somewhat puzzling, therefore, that so many empirical evaluations of the "backwardness" or "progressiveness" of firms choose to ignore the effect of excess capacity at a given production stage on management's attitudes toward an innovation affecting that particular stage.

23. Simon (1973) has referred to these innovations as "metatechnology." The oft-lamented resistance of firms to the implementation of management science models may be explained in part by these considerations.

24. These are discussed at some length in Gold, Peirce, and Rosegger (1970, pp. 230-39).

25. Extensive use has been made in this section of Professor Gold's suggestions.

References

Alchian, A.A., "Uncertainty, Evolution and Economic Theory," *Journal of Political Economy* 58 (June 1950).

Barnett, H.G., *Innovation: the Basis of Cultural Change* (New York: McGraw-Hill, 1953).

Blickwede, D.M., "Development and Implementation of New Technologies in the Steel Industry," in *Proceedings, Third National Seminar: Steel Industry Economics* (New York: American Iron & Steel Institute, 1972).

Capron, W.M., ed., *Technological Change in Regulated Industries* (Washington, D.C.: The Brookings Institution, 1971).

Carlson, J.W., "Aspects of the Diffusion of Technology in the United States," Paper presented for the fifth meeting of the Senior Economic Advisers, United Nations, Economics Commission for Europe, Geneva, October 2, 1967.

Chow, G.C., "Technological Change and the Demand for Computers," *American Economic Review* 57 (December 1967).

Evenson, R., "International Diffusion of Agrarian Technology," *Journal of Economic History* 34 (March 1974).

Gold, B., "Industry Growth Patterns: Theory and Empirical Findings," *Journal of Industrial Economics* 12 (November 1964).

_______, "The Framework of Decision for Major Technological Innovation" in *Values and the Future*, ed. K. Baier, and N. Rescher (New York: The Free Press, 1969).

_______, W.S. Peirce, and G. Rosegger, "Diffusion of Major Technological Innovations in U.S. Iron and Steel Manufacturing," *Journal of Industrial Economics* 18 (July 1970).

Griliches, Z., "Hybrid Corn: An Exploration in the Economics of Technical Change," *Econometrica* 25 (October 1957).

_______, "Profitability vs. Interaction: Another False Dichotomy," *Rural Sociology* 27 (September 1962).

Gruber, W.H., and D.G. Marquis (eds.), *Factors in the Transfer of Technology* (Cambridge, Mass.: MIT Press, 1969).

Hägerstrand, T., *Innovation Diffusion as a Spatial Process* (Chicago: University of Chicago Press, 1967).

Hiestand, D.L., *High Level Manpower and Technological Change in the Steel Industry* (New York: Praeger, 1974).

Ifo-Institut für Wirtschaftsforschung, *Patentwesen und Technischer Fortschritt*, 2 vols. (Göttingen: Otto Schwartz, 1974).

Katz, E., M.L. Levin, and H. Hamilton, "Traditions of Research on the Diffusion of Innovations," *American Sociological Review* (April 1963).

Kennedy, C., and A.P. Thirlwall, "Surveys in Applied Economics: Technical Progress," *Economic Journal* 82 (March 1972).

Keynes, J.M., *The General Theory of Employment, Interest and Money* (New York: Harcourt, Brace, 1936).

Knoerr, A.W., "Role of Literature in the Diffusion of Technological Change," *Special Libraries* 54 (May/June 1963).

Landau, H.G., and A. Rapoport, "Contributions to the Mathematical Theory of Contagion and Spread of Information," *Bulletin of Mathematical Biophysics* 15/173 (1953).

Lynn, F., "An Investigation of the Rate of Development and Diffusion of Technology in our Modern Industrial Society," in *Studies Prepared for the National Commission on Technology, Automation, and Economic Progress*, Appendix Volume II, Technology and the American Economy (Washington, D.C.: U.S. Government Printing Office, 1966).

Mansfield, E., "Intrafirm Rates of Diffusion of an Innovation," *Review of Economics and Statistics*, November 1963.

_______, *Industrial Research and Technological Innovation* (New York: W.W. Norton, 1968).

Nabseth, L., and G.F. Ray, eds., *The Diffusion of New Industrial Processes* (Cambridge: Cambridge University Press, 1974).

Nelson, R.R., "A 'Diffusion' Model of International Productivity Differences in Manufacturing Industry," *American Economic Review* 58 (December 1968).

_______, M.J. Peck, and E. Kalacheck, *Technology, Economic Growth and Public Policy* (Washington, D.C.: The Brookings Institution, 1967).

Peirce, W.S., "Factors Affecting Responsiveness to Technological Innovations in Coal Mining," Working Paper No. 54, Research Program in Industrial Economics, Case Western Reserve University (June 1974).

Rapoport, A., and L.I. Rebhuhn, "On the Mathematical Theory of Rumor Spread," *Bulletin of Mathematical Biophysics*, 14/375 (1952).

_______, "The Diffusion Problem in Mass Behavior," *General Systems Yearbook* (1956). Quoted with permission.

Ray, G.F., "The Diffusion of New Technology: A Study of Ten Processes in Nine Industries," *National Institute Economic Review* 48 (May 1969).

Rogers, E.M., and A.E. Havens, "Adoption of Hybrid Corn: Profitability and the Interaction Effect," *Rural Sociology* 26 (1961).

_______, "Rejoinder to Griliches," *Rural Sociology* 27 (1962).

Rogers, E.M., *The Diffusion of Innovations* (New York: Free Press, 1962).

Rosenberg, N., "Factors Affecting the Diffusion of Technology," *Explorations in Economic History* 10 (Fall 1972).

Ryan, B., and N. Gross, "The Diffusion of Hybrid Seed Corn in Two Iowa Communities," *Rural Sociology* 7 (1943).

Salter, W.E.G., *Productivity and Technical Change* (Cambridge: Cambridge University Press, 1960).

Simon, H.A., "Technology and Environment," *Management Science* 19 (June 1973).

Stigler, G., "The Economics of Information," *Journal of Political Economy* 69 (June 1961).

Taylor, C.T., and Z.A. Silberston, *The Economic Impact of the Patent System* (Cambridge: Cambridge University Press, 1973).

Weinstein, P.A. (ed.), *Featherbedding and Technological Change* (Boston: Heath, 1965).

Part II
Review of Empirical Findings

5 The Sources of Technological Innovations

Myles G. Boylan

This chapter reviews the literature relating to the sources of technological innovations and to the effectiveness of research and development activities in general.[1] Generally, studies relating to actual innovational activities are relatively few, often based on very small samples of data, and offer results that are frequently in conflict.

The terms commonly used in this literature are reviewed in the first section. The measures employed and the availability of data along with their probable impact on models designed to explore relevant major hypotheses are discussed next. Three major hypotheses (and several corollary hypotheses) are presented in the last three sections. They seem to represent the basic foci of the literature covered; its implications and shortcomings are discussed and sample empirical findings bearing on them are reviewed.

Definitions of Terms

Part of the task of explaining the factors involved in the creation of new products and processes is one of *classification*—that is, sorting out the hierarchy of activities that begins with the production of new scientific knowledge and ends with the production of new products (and processes). This task is not simple: Jewkes, Sawers, and Stillerman (1969, p. 25) state:

> There can be no doubt that an over-rigid insistence upon definition would immediately bring all discussion of invention, and the part it plays in changes in ways of living, to a dead stop. The choice must be between discussing these matters with concepts that are necessarily somewhat vague and not discussing them at all.

These authors identify essentially three sequential stages in the process: science, invention, and development. Science is viewed as "directed toward understanding," while the subsequent two activities (called "technology") are "directed towards use" (p. 26). They add that the magnitude of achievement in science can only be judged by other scientists trained in those areas of knowledge, whereas the level of success in technology is ultimately judged by the marketplace.

The distinction between invention and development is one of degree:

"invention is the first confidence that something should work, and the first rough tests that it will, in fact, work" (p. 28). Development is a combination of three phenomena. It is (a) "the stage at which known technical methods are applied to a new problem which . . . has been defined by the original invention(s)"; (b) the stage at which the task to be performed is more precisely defined . . . the search more specific, the chances of final success more susceptible to measurement"; and (c) "the phase in which commercial considerations can be, and indeed must be, more systematically examined . . . " (pp. 28-29). Development encompasses activities concerned with scaling up laboratory models into commercial-sized plants, the search for new and better materials, and the search for new machines capable of operating under greater temperatures, pressures, and tolerances and of processing new shapes.

Ames (1961, p. 369), borrowing from Machlup's ideas, identifies four sequential activities with the output of each stage feeding into the next: basic research, inventive work, development work, and innovation. Basic research draws on the current state of knowledge and uses scientists and other inputs to develop hypotheses, theories, and research papers (formulas), which he labels "new statements about the natural world" (p. 370). Inventive work uses the same kinds of inputs as basic research in addition to the output of basic research to produce "a flow of prototypes of articles which have never been made before or processes that have never been used before" (p. 370). These may or may not be patentable, but generally are working models or sketches. Development work is defined to employ inventions in conjunction with engineers, draftsmen, other labor, materials, power, and fuel to turn out a set of instructions (such as blueprints or diagrams) needed to construct plants with new characteristics of commercial value. Finally, the innovation stage uses the set of instructions to turn the resources of an ordinary construction firm into a new kind of plant. This stage is commercial in the sense that "new" prices appear and "new" transactions take place.

The Ames/Machlup framework is clearly in reasonable conformity with the categories of activities defined by Jewkes, Sawers, and Stillerman (1969). Science is basic research and both use the term "invention" to describe the second stage and "development" to define the third stage. But Ames attempts to develop an air of precision that Jewkes et al. believe is impossible to attain.

The National Science Foundation (NSF) uses a similar framework of activities to define research and development: basic research, applied research, and development (see Sanow, 1959, p. 124). The NSF's coverage may be narrower than that of Ames, Machlup, or Jewkes, however, because the NSF "does not include quality control, routine product testing, market research, sales promotion, sales service, research in the social sciences or psychology, or other nontechnological activities or technical services."

Arrow (1969) has more recently gone to an opposite extreme in classifying these activities. In his view, a typical firm, or individual, is engaged in activities ranging from the production of pure knowledge to the production of pure product. If we were to assign the number 0 to pure knowledge and the number 1 to pure product, then there would be (over any definite period of time) a spectrum of output ranging in characteristics from pure 0 (in very small steps) to pure 1. In short, Arrow prefers not to force the resources employed by the firm, and the firm's subsequent output, into a specific number of categories, such as the four identified by Ames and Machlup. Arrow's view is essentially an extreme interpretation of Jewkes, Sawers, and Stillerman's warning not to attempt strict definitions of the activities underlying technological change.

Occasionally the phrases "inventive activity" and "innovative activity" are employed in the literature. The former refers either to the activity defined by Jewkes, Sawers, and Stillerman, and Ames as invention, or encompasses both science (that is, basic research) and invention. The latter's meaning has to be implied from the context of its usage; it does not usually refer only to the process of innovation as defined by Ames (or to development as defined by Jewkes). In this review, however, we will use innovative activity as synonomous with development.

Research and Development (R&D) is a very common expression in the literature on the sources of technological change. It is used interchangeably to refer both to the *activities* leading up to the production of the knowledge necessary to construct the first commercial facilities designed to produce a new product (or to embody a new process) and to the *inputs* consumed in the production of this knowledge.

Input and Output Measures and Models of Technological Process

The above activity groupings may now be compared with the measures of inputs and outputs found in the literature. In each activity, it is possible (in theory) to measure the cost of inputs consumed with reasonable accuracy, except for the freely available scientific and technical knowledge that is part of a growing international legacy. For example, in the second stage activity (invention or applied research), the input costs would cover expenditures on basic research, purchased new knowledge from outside groups, and tangible resources consumed in the production of new knowledge and in the construction of prototypes of new products and processes within the organization. In measuring expenditures on basic research, as Kuznets (1962) and Jewkes, Sawers, and Stillerman (1969) point out, it may be difficult to identify the resources that were consumed solely in the production of that sector of new knowledge which served as an input to the applied research activity involved in the development of a particular invention, because basic research activities are generally not focused on particular goals. On the other hand, as Mansfield (1968, p. 59) and others have noted, basic research amounted to only 4 percent of total R&D expenditures in the mid-1960s in manufacturing.

It is extremely difficult, however, to measure the output of basic research activities accurately, whatever the locale or level of aggregation. Such output—new scientific knowledge—does not have immediate commercial value, nor is it produced in homogeneous units. Even a group of expert scientists is unlikely to reach a consensus about the relative significance of new findings.

Schmookler (1952, 1966) and Scherer (1965b), among others, have used patents as a measure of the output of applied research, while readily admitting the shortcomings of this measure. For example, Sanders stated that two basic conditions should be satisfied before patents are used as a measure. First, both the proportion of inventive activity leading to results that are legally patentable, and the fraction of such results that actually are patented, should remain invariant over time. Sanders (1962) offered seven reasons why this condition is not fulfilled. Second, the quality of patents (which he did not define, but which could be defined to be the net present value of their total future commerical value) should be uniform on average from year to year, at least for particular process or product areas such as steelmaking. This condition is clearly not fulfilled. (See Sanders, 1962, pp. 68-70; Kennedy and Thirlwall, 1972; and Mansfield, 1968, p. 34). An alternative measure is to "count" the number of significant inventions developed in a given sector of the economy. This is done on an *ex post* basis after the inventions have been turned into innovations and become widely diffused throughout the economy. Jewkes, Sawers, and Stillerman (1969), Hamberg (1963), and Mansfield (1965) have used this approach. But "significance" does not imply homogeneity. More importantly, the accumulation of many small inventive advances often outweighs the effects of major inventions. Jewkes et al. credit Schmookler with this important observation and are in complete agreement with it (p. 209).

The output of the third activity—the development of innovations—is also extremely difficult to quantify. One approach has been to "count" the significant innovations developed in a given sector of the economy. This method is subject to the same limitations as counting significant inventions. Another highly controversial method has been to use the stream of "profits" earned from the innovations during the diffusion process (that is, the increase in profitability on old product lines resulting from new processes, or the profits earned from the sale of new products) as a measure of output. But the immediate implication is that accurate measurement will often have to encompass a broad expanse of time after the innovation has been developed to capture all returns from it. Thus the measure is not really homogeneous because adjustments must be made for the time value of money. An even greater source of variability in this measure concerns the problem of adjusting for reduced profits in old facilities and old products as the innovation diffuses. And more fundamentally, what is the basis for believing that new products or processes will be able to retain greater earnings than those yielded by old products and processes in the past, even after an increasing number of producers adopt the innovations? It must be stressed that those employing this or related measures (such as Griliches, 1958, 1964; Mansfield, 1965, 1968; and Minasian, 1962, 1969) do so in a very crude manner, involving an approach very similar to the attempts of others to measure the rate of technical progress with an aggregate production function (as reviewed by Baird in Chapter 1).

For example, Minasian (1962, pp. 102-111) used the increase in total factor productivity as a measure of R&D output. He used value added deflated by a carefully constructed wholesale price index as a measure of a firm's output, employed labor expenditures deflated by a wage rate index as his measure of labor, and gross plant and equipment value deflated by the GNP implicit price deflators for new construction and durable equipment as his measure of capital. Total real input was estimated by combining capital (K) and labor (L) into a weighted average via the Cobb-Douglas production function:

$$\text{Total Input}_t = L_t^a \cdot K_t^{1-a} \tag{5.1}$$

$$\text{where } a = \frac{\text{wage bill}}{\text{wage bill} + \text{profits before taxes} + \text{interest paid} + \text{depreciation}}$$

Finally, a productivity index was constructed

$$P_t = \frac{\text{VA}_t}{\text{Total Input}_t} \tag{5.2}$$

for each year from 1947 to 1957. The annual change in the ratio P_t/P_{1947} from 1947 to 1957 was regressed against e^{rt} (where t is time and r is the annual growth rate) to get an estimate of the average annual growth (r) in productivity. This was also done for shorter periods (such as 1952-1957). It was this growth in productivity that Minasian hypothesized could be caused by R&D efforts. This was the "output" of R&D activities.

In a more direct approach, Mansfield (1968) and Minasian (1969) have used a broadened Cobb-Douglas production function to estimate the marginal returns to R&D expenditures, for example:

$$\text{VA}_t = A \cdot L_t^a \cdot K_t^b \cdot (\text{R\&D})_t^c \tag{5.3}$$

where A is a constant and $(\text{R\&D})_t$ is the cumulative expenditures over the previous n years. This allows them to calculate the marginal product of R&D expenditures at time t as $c \times \text{VA}_t/(\text{R\&D})_t$ (where VA and R&D are the average value for all firms or industries in the sample). Neither Mansfield nor Minasian claims the results to be anything but a crude approximation of reality

Comanor (1965) and Comanor and Scherer (1969) have used the sales volume of new products in years immediately following their introduction as another measure of innovative output. This measure has obvious disadvantages. Lacking adequate measures of the output of basic research, applied research, and development, little effort has been made to model these activities separately. It is simply not possible to establish quantitative links between their respective inputs and outputs, as has been emphasized by Sanders (1962).

Most studies have concentrated on one or more of three limited objectives. Perhaps the most common objective has involved efforts merely to measure inventive-innovative activity, either absolutely or relative to all other economic activity. One method measures R&D inputs in terms of expenditures reported by firms or in the *National Science Foundation Reviews of Data on Science Resources* (see Mansfield, 1968, pp. 44-54). In this connection, both Sanders (1962, p. 59) and Mansfield (p. 53) have noted that there is likely to be an upward bias in reported figures over time because of amendments to the 1954 Revenue Act that allow firms to deduct R&D expenses before taxation. Mansfield also notes (p. 53) that firms have difficulty in separating development from production expenditures (and applied research from development expenditures). Sanders (pp. 58-59) has stated the NSF figures are incomplete because they omit private individuals' and some universities' inventive efforts. Others have measured R&D inputs in terms of the number of R&D employees: for example, Maclaurin (1954) used the number of employees with Ph.D.s; and Schmookler (1954) used the number of technical employees. Kamien and Schwartz (1975, p. 5) point out that each possible measure of employees is subject to a number of deficiencies. The basic problem is that R&D-type activities may be carried out in operating or other divisions of the firm, hence the number of designated R&D employees alone underestimates R&D activities, whereas the number of all scientific and technical personnel tends to overestimate them. Also, there is little homogeneity of R&D employees among firms and over time, as Sanders (1962, p. 57) has noted.

Efforts to measure the inputs of R&D activities includes the use of patent statistics; see, for example, Schmookler (1966) and Comanor and Scherer (1969). But many studies have concluded that patent statistics are only a crude measure of inventive-innovative activity for reasons previously discussed. Also, Schmookler (1966) and Sanders (1962) have expressed the belief that there has been a drop-off in the fraction of inventive output being patented. Two other methods have centered on counting the number of "significant" inventions or innovations, or measuring the growth of total factor productivity, both subject to weaknesses which have been discussed previously.

A second group of published studies have sought to reach beyond the sheer measurement of R&D inputs and outputs by searching for relationships between various stages of the development and utilization of technological advances. For example, in the study partly reviewed above, Minasian (1969) found strong correlations between R&D expenditures and productivity growth a few years later in chemical firms. Schmookler (1966) found relative changes in the numbers of scientists and engineers to be positively correlated with changes in the number of patents in a sample of industries during the 1870-1950 period. Mansfield (1968) reported significant positive correlations between the number of "important" inventions developed by firms in the chemical, petroleum and steel industries and their R&D expenditures, after adjusting for the size of the firm. Comanor and Scherer (1969) found the number of R&D personnel was significantly correlated with the number of patents received and the sales volume of new products (within two years after their introduction) for fifty-seven drug companies, when firm size was held constant. But it is apparent that this sample of findings cannot be generalized at the aggregate level, as will be discussed in greater detail later.

A third group of studies have sought to relate the above measures of R&D activities to firm size and other aspects of market structure to examine interrelationships between technical progress and market structure. These will be discussed in the last section.

The conclusion of this section must be basically negative. The output measures have been found to be uniformly weak. R&D expenditures provide a somewhat better measure of input but are still subject to substantial inaccuracies. R&D employment is apparently quite weak, because inventive-innovative output is likely to be strongly related to the quality of R&D personnel, and, as noted by Sanders (1962, p. 59), a mere counting of heads does not measure this elusive quality. In view of the serious and pervasive weaknesses of the measures employed, findings of positive correlations among some of them in isolated studies yield but little assurance. How many studies have not been accepted for publication, or even submitted, because no significant correlations were found?

Finally, our apparent inability to penetrate the three basic activities in terms of effective measurement suggests that, when working with the available data, we are primarily studying the development stage of technological improvement processes. In the mid-1960s, according to the *National Science Foundation Reviews of Data on Science Resources* (No. 7, 1966), 76 percent of R&D expenditures in all manufacturing represented development expenditures. In this sense, both R&D expenditures (or employees) and patents (or significant inventions) represent primarily complementary inputs into development work, although patentable findings can also, and often do, occur as byproducts of developmental efforts. Accordingly, the foregoing review of the literature provides no answer to the very important question of whether there are empirically demonstrable determinants of the effectiveness with which scientific research is organized, its targets chosen, its progress accelerated, and its commercially valuable outputs increased.

Hypothesis I: Individual versus Team Contributions to R&D

One of the most pervasive hypotheses relating to the sources of technological change suggests that *the pace of R&D expenditures has been accelerated by a shift from individual efforts to larger organized research teams in industrial, university and government laboratories.* Its corollary would claim that *organized teams in large R&D laboratories have been displacing individual inventors as the major source of inventions.*

This hypothesis (and its corollary) underlies at least part of the work of Jewkes, Sawers and Stillerman (1968), Carter and Williams (1957, 1958), Hamberg (1963), Mansfield (1968), and many others. Some support for the hypothesis is provided by the growth in the number of commercial research laboratories from a mere handful at the turn of the century to 300 in 1920, over 5,400 in 1960 (Silk, 1960) and over 13,000 in 1964 (Mansfield, 1968). R&D expenditures by firms, universities, and government laboratories have also accelerated, both absolutely and relative to GNP. But we do not have accurate statistics on the number of part- and full-time individual inventors. Total expenditures on R&D rose from less than \$1 billion in 1940 to \$3 billion in 1950, \$13 billion in 1960, and \$20 billion in 1965, according to the NSF, again excluding private inventors and some university expenditures. Markham (1965) indicates that R&D expenditures from industry and government sources as a percentage of GNP rose from 0.15 percent in 1930 to 1.16 percent in 1950, and to 2.41 percent in 1959. R&D expenditures originating from industry sources alone amounted to 0.13 percent of GNP in 1930, 0.53 percent of GNP in 1950, and 0.93 percent of GNP in 1959, showing the same pattern (Markham, 1965, p. 76). Finally, 80 percent of the patents issued at the turn of the century were issued to individual inventors (and not immediately assigned), while only 30 percent of the patents granted in 1970 were issued to individuals and not immediately assigned. (See Kennedy and Thirlwall, 1972.) Although these figures do not indicate the difference in the quality of patents assigned to corporations versus those assigned to individual inventors, the implications of these statistics seem to confirm the hypothesis.

There is conflicting evidence about the corollary, however. Jewkes, Sawers, and Stillerman (1969) studied sixty-one significant twentieth-century inventions and found that individual inventors were primarily responsible for the development of thirty-three, while only twelve originated in large part from large R&D laboratories. Individual inventors were found to play the most important roles in the first two stages, basic research and applied research, whereas corporate laboratories were more important in the development process. These "bare" quantitative findings were more substantially supported by carefully researched case studies of these inventions from their origins in basic research, where it proved possible to make this determination, through the development process. In many of these cases, it was the work of individuals scattered geographically and often working years apart, sometimes aware and at other times unaware of the work of others in the same field, that brought the research close to the development stage. In several other cases, individual inventors were invited to continue their research in corporate laboratories and sometimes were requested to aid in the development process. The impression gained from these studies is that, even when the bulk of the R&D effort was carried out in large laboratories, it was the insight of a few men that led to the necessary breakthroughs. That is, the corporations contributed economic support, often after creative individuals had contributed the needed originality. Jewkes, Sawers, and Stillerman have expressed serious doubts about man's ability, as yet, to organize the production of originality.

Further support for the role of individual inventors is provided by Hamberg's finding (1963) that, of twenty-seven major inventions developed from 1946 to 1955, twelve came from individual inventors and only seven from large R&D laboratories. Enos (1962) identified fourteen important process inventions in the petroleum refining industry covering the twentieth century (three of which might be labeled as the development stages of three other inventions included in this group), of which ten were clearly the work of one or two inventors (two of the remaining four inventions might normally be called the development stage of previous inventions). In four of these ten inventions, the inventors also contributed to the development work.

Opposite evidence was uncovered by Peck (1962) who studied 194 inventions in the 1946-1957 period covering four technical areas of the aluminum industry. He found independent inventors to be an unimportant source of these inventions, although it should be noted that most of these inventions are hardly significant relative to those covered by Jewkes, Hamberg, and even Enos. Thus, although the available literature covers only a small sample of important inventions during the twentieth century, the results so far indicate that the corollary has by no means been substantiated. It will take many more micro studies of such types to clarify the respective roles of individual inventors and large corporate laboratories as well as their relative contributions and efficiency.

Hypothesis II: Inventive-Innovative Efforts and Economic Benefits

Perhaps the single most important and most widely accepted hypothesis in this area is that *increases in the volume of inventive and innovative efforts have led to attractive economic benefits to those undertaking them.* Its corollary would be that *the return to investment in R&D activities by firms has been very high.*

Empirical Findings

Studies investigating this hypothesis and its corollary can be crudely classified into two groups. One attempts to show a significant positive correlation between R&D inputs (measured as R&D personnel or expenditures) and R&D output (measured as patents, number of "significant" inventions or innovations, or relative sales volume of new products some years after their introduction). The other explores the connection between R&D inputs (usually measured by expenditures) and R&D benefits (measured as increases in total factor productivity or additions to profit).

Illustrative of the first group are several studies reviewed in previous sections. Schmookler (1966) found significant and positive correlation between R&D employment and the number of patents issued between 1870 and 1950. Also, Mansfield (1968) found R&D expenditures by firms to be significantly and positively correlated with the number of their important inventions in the chemicals, steel, and petroleum refining industries in the 1946-1962 period, after adjusting for firm size. Along similar lines, Comanor and Scherer (1969) discovered significant and positive correlations between the number of R&D personnel in a sample of fifty-seven drug firms and both the number of patents issued to them and the sales volume of new products two years after their introduction, after adjusting for size of firm.

In addition to this representative sample, Pavitt and Wald (1971) found a strong positive correspondence between the research intensity (measured by R&D expenditures as a percentage of sales) in thirteen industry groups[2] and the percentage of their sales in 1969 that was expected (by member firms) to be accounted for by products introduced after 1966.

Though the preceding four studies support the hypothesis to some extent, they are subject to the significant weaknesses discussed at greater length earlier. For example, they represent only a very limited proportion of firms within the given industries, and there seems to have been a marked tendency to bias the sample in the direction of large firms with large and successful R&D laboratories. Available studies also provided a very limited coverage of industries. Chemicals, pharmaceuticals, and petroleum refineries have accounted for most investigations. These industries have been analyzed unevenly over a dispersed array of time periods. Closer scrutiny also reveals the use of questionable modeling techniques, particularly in selecting the time lags between R&D expenditure and R&D "output" to measure the "causal" effect implied by the interpretation of results. All these sources of vulnerability are in addition, of course, to the very imperfect measures of input and output employed. Further doubts stem from the likelihood that published studies include a large percentage of the successful attempts to correlate R&D inputs and outputs relative to the proportion of unsuccessful attempts likely to achieve publication. Despite our misgivings about the widespread applicability of such findings, Kamien and Schwartz (1975, pp.

5-6) stated: "There seems little doubt that, on average, a direct relation between innovational effort and innovational output exists. However, it is likewise true that the transformation may depend on factors other than effort and it may not be linear." In all likelihood, however, their conclusion more accurately represents the *beliefs* of those who have been involved in assessing the effectiveness of R&D efforts than convincingly generalizable findings. Accordingly, we prefer not to accept the hypothesis until more persuasive evidence has accumulated.

Few published studies can clearly be classified into the second group relating R&D inputs to benefits beyond output and sales. Minasian's study (1962), parts of which were outlined earlier in this paper, was one of the early attempts to link R&D inputs with an indirect measure of commercial benefits: total factor productivity.[3] His study was based on data supplied by twenty-five chemical and pharmaceutical firms, covering the eleven-year period from 1947 to 1957. Measures of inputs and output (that is, total factor productivity) were defined over broad periods: six- to eleven-year intervals. R&D expenditures were measured relative to productive inputs and the latter were measured either as a geometrically weighted average of capital and labor[4] or by capital alone. The basic design of his model was to search for the highest correlations between total factor productivity (the dependent variable) and the ratio of R&D expenditures to expenditures on other inputs (the independent variable).

This search was accomplished by systematically varying the intervals over which total factor productivity and relative R&D expenditures were measured from six to eleven years, and by varying the number of years by which the starting date of the interval in which relative R&D expenditures were measured preceded the starting date of the interval used to measure average total factor productivity from zero to five years.[5] The length of the intervals made a great deal of difference. Variations in relative R&D expenditures explained from 53 to 70 percent of the variance in total factor productivity when productivity was fitted over nine- to eleven-year intervals and R&D expenditures were averaged over nine- to ten-year intervals. Shortening the intervals over which productivity and R&D expenditures were averaged (to as low as six years) progressively lowered the explained variance in productivity to 16 to 27 percent. The length of the lag period made little difference in all cases.

Employing other data covering profitability (that is, the rate of return on invested capital), the relative capital intensiveness of plant operations, and the average size of plant, Minasian also demonstrated in his framework that (a) there was practically no correlation between changes in total factor productivity and changes in either capital intensity or size of plant, suggesting that the link between relative R&D expenditures and future total factor productivity was not the indirect result of strong correlations between R&D and capital intensity or plant size; and (b) there was practically no correlation between changes in relative R&D expenditures and earlier profitability, suggesting that the correct order of causality was R&D-affected productivity and profitability, rather than vice versa.

Several important questions are left unanswered by this study. First, if the correct order of cause and effect runs from R&D to productivity and profitability, why did Minasian neglect to correlate R&D directly with future profitability? We suspect the reason is because such correlations proved to be low; that is, that the correlations between total factor productivity and concurrent profitability was low.[6] Next, no reason is offered to justify analytically the high correlations between total factor productivity and relative R&D expenditures when both these variables are averaged over nine- to eleven-year periods (as opposed to the much lower correlations associated with six- and seven-year averages), with almost no time allowed between cause (R&D) and effect (improved productivity). Indeed, the maximum lag of two years may not even be enough time to incorporate some of the process innovations into existing plants, even assuming instantaneous development and no start-up problems. Finally, Minasian's ad hoc decision to employ relative rather than absolute R&D expenditures, or some combination of the two, is questionable. In short, Minasian's approach is permeated with methodological convenience; that is, it maximizes correlations that superficially imply significance, but lacks analytical integrity. In a comment on Minasian's study, C. Hitch stated: "Minasian succeeds in 'explaining' inter-firm differences in profits in one industry by differential research activity; others have made similar attempts in other industries and failed" (1962, p. 193).

A later study by Minasian (1969), based on the same sample as the above study, investigated the relative returns to investment in capital and investment in R&D. Using a Cobb-Douglas production function with capital, labor and cumulative R&D expenditure as inputs, and value added as output (see the earlier section on measures and models) he found:

$$VA_t = A \cdot L^{0.82} K^{0.156} \left\{ \sum_{1947}^{1957} \text{R\&D} \right\}^{0.113} \tag{5.4}$$

where $R^2 = 0.99$

Using the technique outlined previously, he found that an increase of $1.00 in the gross stock of plant and equipment (K) would lead to a $0.09 annual increase in value added (over the indefinite future), whereas the same dollar increase in cumulative R&D expenditures would result in a $0.54 annual increase in value added, which supports the corollary to the hypothesis.

Using a methodologically similar approach, albeit with minor technical differences, Mansfield (1968) studied the effectiveness of R&D in ten industry groups[7] in the 1946-1962 period, and found industry rates of return to additional R&D expenditures to be 15 percent or more. In the same study, he also estimated the returns to R&D expenditures by five large chemical firms and five large petroleum refining firms in the 1946-1962 period. By varying his assumption about the nature of technical progress (embodied or disembodied), he obtained rates of return as high as 60 percent for petroleum refining, assuming embodied technical progress, to a low of 7 percent for chemicals assuming disembodied technical progress. On average, Mansfield's estimates are considerably below Minasian's 54 percent but are still high (see Chapter 2). One difficulty both Minasian and Mansfield faced in their firm level studies were nonrandom samples of data. Thus, their results are noncomparable.

In two interesting nonmanufacturing studies, Griliches (1958 and 1964) concentrated on the returns to a particular innovation affecting only one basic product: hybrid corn, which was first used commercially in 1933 and which diffused rapidly during the late 1930s. This is significant because cause and effect can be established with greater precision. Using questionnaires to obtain records of expenditures on R&D in hybrid seed corn by state agricultural experiment stations, private seed companies, and the U.S. Department of Agriculture over the 1910-1955 period, Griliches obtained a high enough response rate to feel justified in extrapolating probable R&D expenditures by nonrespondents. These expenditures were then summed into a single future value by "carrying forward" pre-1955 expenditures at a 5 percent annual rate of interest (10 percent was also tried—with little effect on the results). Griliches measured the "gross social returns" of these expenditures by the following means. First, he estimated the increase in corn yield per acre valued at market prices obtained by farms adopting this new kind of seed to be 15 percent. Then, he estimated the gain in consumer "surplus" resulting from the drop in corn prices that would have occurred if the acreage devoted to corn growing had remained the same, as if this innovation had never materialized. A reasonably accurate measure of the price elasticity of demand for corn was available (0.5) to facilitate the second estimate, whereas the elasticity of supply would have had little effect on the estimate. The additional cost of purchasing hybrid seed was subtracted from gross returns to get an estimate of net social returns. The net returns were subsequently summed into a single future value by carrying them forward to 1955 at a 5 percent annual rate of interest. Treating the summed value of R&D expenditures as a single investment, the annual net social rate of return was estimated to be 700 percent, a figure which Griliches considered to be conservative, although he admitted his estimating procedure was subject to a wide margin of error. This estimate does not include the benefits enjoyed by the private seed companies that was reflected in the higher price of hybrid seed. Accordingly, Griliches measured only the "free" net benefits accruing to farmers adopting this innovation (which may have been negative in years in which corn prices were below its average production cost) and to consumers of corn. The cost of "manufacturing" hybrid seed was not stated; but this seed sold for a price approximately six times higher than conventional corn seed in a "typical" year. Thus, it is possible that the direct benefits to seed manufacturers from their cumulative R&D expenditures (and the expenditures by government research groups) were also quite high. In a follow-up study (1964), Griliches estimated the net social rate of return to be at least 1000 percent.

In comparison with the Mansfield and Minasian studies, Griliches's estimates of R&D expenditures are probably at least as accurate, and his linkage between expenditures and R&D benefits is probably far more accurate. His estimates of social benefits, however, were obtained by treading the same dangerous methodological grounds. Although farmers adopting this innovation in the 1930s undoubtedly gained from the increased yield per acre (because corn prices were still dominated by the productivity or cost of growing corn from conventional seed) these benefits must have nearly disappeared by the mid-1940s when the use of hybrid seed was standard practice, *unless* a stream of improvements in hybrid seed were made throughout this period (which is not clear). Though these benefits would normally be passed on to consumers of corn (and corn products) in the form of lower prices as the innovation diffused, possible overproduction and resultant government price supports by no means lend assurance that benefits *were* passed on to consumers. Nor did he take account of the cost of the considerable efforts made by the Department of Agriculture's Extension Service to promote the shift to hybrid seed corn. Nevertheless, it is noteworthy that these two published studies do support the corollary to the hypotheses.

In their review article, Kamien and Schwartz (1975) stated: "despite methodological difficulties and debates, all the evidence—at the level of the firm, industry, and economy—indicates that the contribution of R&D to economic growth/productivity is positive, significant and high. Estimates of rates of return range between 10 percent and 50 percent, with a bias toward the higher values." This conclusion may reasonably be regarded as unduly optimistic, however, for two reasons.

First, at the industry level, we still are heavily dependent on Mansfield's investigation of ten industry groups in which R&D input was linked to product output via a Cobb-Douglas production function format, with all the attendant difficulties and conceptual biases of that approach. And even if we assume that his data were accurate, there would still be the problem of lag structure to be dealt with as well as the question of statistical significance, since his observations were few (covering only seventeen years). Mansfield himself considered his findings to be only "tentative" (1968, p. 33), and not proving a direction of causation that runs from R&D expenditure to growth (rather than vice versa) because he was not able to cope with the proper specification of lag structure. Griliches's study of the hybrid corn industry is atypical in that it studied only one (successful) result of government and private R&D efforts in agriculture and hence cannot answer the question of the overall effectiveness of governmental and private R&D efforts in agriculture.

Second, at the level of the firm, published estimates of the returns to R&D expenditures are uniformly high. But the coverage of these findings is limited to a few firms in a few industries, especially those recognized as having had successful R&D programs, such as chemicals and pharmaceuticals. Moreover, the findings are based on nonrandom samples of firms, including few studies of firms with poor R&D records; on voluntarily provided questionnaire data, which are less likely to provide unfavorable results; and on (in part) ad hoc methodology, such as that found in Minasian's study (1962).

Thus, it does not seem unreasonable to conclude that the fragmentary studies available as yet have been insufficient to test this hypothesis and its corollary. Indeed, one need only ask what proportion of total R&D expenditures has been covered by studies attempting to trace the specific effects of actual resulting innovations to recognize the extreme limitations of the exploratory probings and speculative inferences which have been published so far.

Difficulties in Confirming this Hypothesis

A review of the literature concerned with linkages between R&D inputs and results reveals three major difficulties besetting empirical attempts to confirm the hypothesis and its corollary. The first is the lack of comprehensive and accurate data, which has already been discussed. The second derives from the difficulties involved in seeking even to conceptualize, let alone measure, R&D outputs. Suppose one were able to measure for a firm or industry the movement of total factor productivity, or the value added, or the sales or the profits earned by new products or processes over some specified period following their introduction. Clearly none of these would have measured the output or benefits of R&D activity with precision, for the following reasons: (a) a change in the price of inputs or a change in their quality, possibly induced by R&D efforts in supplying industries, would affect all such measures; (b) comparably effective R&D efforts by competitors would tend to reduce benefits to the firm or the industry in a competitive environment; and (c) changes in demand, such as demand fluctuations over a business cycle, would affect all such measures. Simply stated, the effects of factors not directly related to the firm or industry's internal R&D efforts will affect all the methods employed to measure the output of such R&D efforts in empirical research.

The third set of difficulties is reflected in the variance among innovations in the time that elapses from the initiation of a well-defined R&D project to its fruitions in the form of commercial production and sale of the results, and subsequent variations in rates of diffusion among all producers as well as additional variations in output and capacity utilization rates. These considerations are necessary to construct models in which the average elapsed time between R&D expenditures and R&D benefits is accurately specified. Yet, as seems likely, even the average time lag for a major firm with a large R&D laboratory engaged in many different R&D activities at any specific time is likely to vary substantially among innovations and from one year to the next. By way of illustration, one need only consider the difference in such gestation periods between innovations requiring the construction of major new facilities as compared with others involving only minor modifications in processing and product design. Furthermore, neither the calculation of an average lag nor the interpretation of such an average is obvious. Indeed, it is not clear that such data for highly variegated developments can be usefully combined.

Reports on R&D Benefits from External Innovations. Brozen (1965) recognized the unevenness in the intensity of R&D efforts across industries, some industries serving as net suppliers of innovations, others as net receivers. And the same observation obviously applies to firms within any given industry. Accordingly, Brozen concluded that the Schumpeterian hypothesis that firms must be big to compete effectively via innovations was false. Net receivers of innovations might have to be big for other reasons (such as economies of scale in capital facilities) but not to benefit from the R&D output of others. His conclusions were based on general observations, however, rather than on a detailed analysis of specific innovations.

Mansfield (1968, p. 67) recognized that the estimated return on R&D expenditures in a specific industry or single firm will often not include benefits received by using firms. It is apparent, of course, that by licensing innovations or by purchasing equipment which embodies innovations, firms may enjoy total innovation benefits that exceed those traceable to their own R&D expenditures.

Peck (1962) studied 194 inventions in the post-World War II period in the aluminum industry in four technical areas (joining, finishing, fabricating, and alloy development). Primary aluminum producers accounted for only 11 percent of the innovations in the first three technical areas, whereas equipment manufacturers accounted for 50 percent and other industries (related in some way to primary producers) accounted for the rest. Peck noted that at least 40 percent of the output of primary producers during the late 1950s went into products made possible by these innovations and that the proportion was likely to rise during the 1960s as these innovations became more widely diffused. Thus, even though primary producers accounted for more than 75 percent of the innovations in the fourth technical area (alloy development), it would prove difficult to accurately link all R&D expenditures by primary producers with an effective measure of all benefits resulting from these expenditures.

Mueller's study (1962) of DuPont's twenty-five major product and process innovations during the 1920-1950 period reached similar conclusions. Ten of these were developed in DuPont's highly respected research department, though fifteen came from outside sources and were comparable in importance to those developed by DuPont. DuPont was relatively more successful in process and product *improvement* innovations, accounting for five out of seven. But only five of eighteen new products came from DuPont's laboratories, indicating the importance of outside firms (or individuals) in this area. Mueller also believed that "the quantity of [less spectacular innovations] would be found to be more closely correlated with the size of DuPont's research expenditures during 1920-1950" (p. 345). But only speculative evidence was offered on this last point, in part reflecting difficulties in linking expenditures with innovations, as previously discussed.

Peck and Mueller are in close conformity on the following two points: (a) they found the relative contributions of a firm or industry to innovations of direct value to it to be high when the innovations were of the product improvement type, or when they were necessarily based on some operating experience; and (b) they found the relative importance of outside firms to be much greater in supplying innovations that were much less related (or not related at all) to immediate production and product knowledge.

Reports on Lags between R&D Expenditures and Benefits. Mansfield (1968) recognized the importance of correctly specifying lag structures in his study of ten major chemical and petroleum refining firms, but reported he was unable to obtain data that would allow him to estimate the lag between R&D expenditures and increased value-added in new products or processes. Consequently, he could not be certain whether increased R&D expenditure was responsible for such increases in the value-added (in his Cobb-Douglas production function format) that were not caused by a change in the volume of capital and labor services, or vice versa.

Enos (1962) carefully studied the lags from the end of the invention stage (when formulas and working models indicated that a new process had good technical and commercial probabilities of success) to the end of the development stage (when the first commercial plant went into operation) for nine process innovations in the petroleum refining industry. He found that such lags ranged from 5 to 24 years and that seven had lags from 8 to 13 years. In the same study, Enos found that of thirty-five other product and process innovations in areas of mechanical, chemical, pharmaceutical and electronic technology, twenty-nine experienced lags distributed fairly uniformly over periods ranging from 1 to 14 years, three had lags from 22 to 27 years, and three had lags from 53 to 79 years. Perhaps of even greater significance are the results of Lynn's study of twenty major innovations whose commercial development was started in the 1885-1950 period (1966). By carving this period into three subperiods (1885-1919, 1920-1944, 1945-1964), Lynn found the average lag from the end of the invention stage to the end of the development stage was 37 years in the earlier period, 24 years in the middle period, and 14 years in the post-World War II period. Applying the same analysis to the Enos sample, we found that twenty-eight (out of thirty-five) inventions were produced between 1884 and 1945. The twelve produced between 1884 and 1919 had an average lag of 16 years, and the sixteen produced between 1920 and 1945 had an average lag of 7 years. Lynn also found the total reduction in lag times resulted from a reduction in incubation period (defined as the length of time between the end of the invention stage and the beginning of commercial development) which fell from 30 to 16 to 9 years on an average. Thus, one effect of increased R&D effort may have been to accelerate the process whereby inventions are recognized to have commercial value.

The major conclusion of this subsection is that the significant variations in elapsed time from the end of the invention stage (defined as before) to the end of the development stage recorded for some individual innovations, combined with the clearly implied existence of similar variations within the applied research stage and within the diffusion period, represent very substantial barriers to effectively linking the *total* annual R&D expenditures of firms or industries with subsequent streams of R&D benefits on any generalized basis.

On the Pace of Inventive-Innovative-Outout: Alternative Theories

Because of the enormous difficulties confronting attempts to substantiate the hypothesis and its corollary empirically, it is not surprising that a number of theories about the process of invention have been proposed. In this concluding subsection, such efforts to classify prevailing beliefs about the pace of invention and its determinants will be reviewed and contrasted with the hypothesis.

Among the theories which seem to support the hypothesis, Usher identified a "cumulative synthesis" school and Jewkes, Sawers, and Stillerman noted the existence of a "doctrine of exponential growth" in inventions. The underlying idea is that each invention opens up the way for two or more second generation inventions, etc., leading to a geometric growth rate. The exponential growth in measured R&D expenditures is seen to reflect this school of thought. There is very little empirical support for this theory—even less than for the hypothesis—because no attempt is made to establish a sequence of connections and a direction of causality that runs from R&D inputs to R&D output.

Usher also identified two "schools of thought" that may be regarded as neutral with respect to the hypothesis. One was labeled the "transcendentalist school" because it specifies that inventions cannot be planned or predicted but rather are the product of inspiration or genius. A second school of thought—or theory—is called the "mechanistic school" because possible economic motivations underlying the development of inventions are given major emphasis. Kennedy and Thirlwall (1972) have noted that these two schools of thought are rivals unless inventors are primarily motivated by economic gain. Mansfield interprets the mechanistic school's theory to imply: "if the great inventive geniuses had never lived, the same inventions would have been made by others without serious delay" (1968, p. 51). From this viewpoint, these two theories are certainly rivals.

A third group of theories represent rivals to the hypothesis. For example, Jewkes, Sawers, and Stillerman claimed that some students of technological change believe all easy inventions have been made, and, hence, that the flow of inventions and innovations may well decline in the future. Machlup (1962) has supplied an underlying framework for such beliefs. Assuming the most important input into innovative activities is scientific and technical labor, he concluded that: (a) the law of diminishing returns applies to attempts to increase the number of trained personnel via increases in salaries; (b) the average quality (that is, inventive capacity) of the labor force tends to decline because those with the greatest innate ability in inventing are more strongly motivated by factors that go beyond financial compensation; (c) as the level of inventive employment increases (adjusted for quality differences), the amount of new inventions would not increase as rapidly; and (d) as the number of new inventions increases, the number of effective inventions would not increase as rapidly (that is, the average quality of inventions would fall). Machlup concluded that these four sources of diminishing effectiveness of accelerated R&D effort "may add up with a vengeance" (p. 167). Machlup's four sources of diminishing returns may apply in the short run when massive efforts are made to accelerate the rate of effective inventions. But there is no evidence whatever to support any of his assumptions over the longer run.

Hypothesis III: Economies of Scale in R&D Operations

Another area of analysis which has attracted increasing attention in recent years concerns the hypothesis that *there are economies of scale in the operation of large R&D laboratories.* One corollary would be that *large firms have relatively larger and more effective R&D operations than smaller firms.* This suggests that larger firms spend at least the same proportion of sales revenue on R&D as small firms, and that the resulting larger R&D operations yield more than proportionately greater benefits, either because larger expenditures can attract better personnel or because larger staffs are more likely to include scarce creative talents. Another partial corollary would suggest that *more diversified firms would benefit more than narrowly specialized firms from R&D operations because the products of such operations often range beyond the initial target area and, hence, can be more fully utilized by more diversified firms.* On the other hand, it has also been argued that more diversified firms have to spread limited R&D budgets over a wide range of activities, thus limiting the prospect for greater advances in selected technical areas.

Another partial corollary of the above hypothesis is that *firms in highly concentrated industries spend a larger proportion of their sales revenues on R&D activities than firms in less highly concentrated industries.* This suggests that concentration may have increased in some industries because firms with proportionately larger R&D operations achieved a string of innovation successes which subsequently increased the level of concentration. On the other hand, it has been argued that this corollary will not be supported as a generalization unless separate account is taken of significant interindustry differences in such areas as technological opportunities, product differentiation and the level of buyers' technical expertise, and barriers to entry by new producers.

All hypotheses relating R&D programs to firm size and market structure imply increasing returns to scale in R&D activities. For example, Hirsch (1969, p. 41) stated: "scale effects in the R&D process appear to be highly significant—especially for the kind of product and production technique development programs that are associated with the technologically progressive firms." Fisher and Temin (1973, p. 58) interpret this hypothesis and its first two corollaries as the basic proposition of Schumpeter and claim there are three parts to it. First, there are returns to the scale of R&D staff because of the greater specialization of personnel (the main hypothesis). Second, increases in the size of firm, given the scale of R&D, will lead to greater economy of use (of the staff and its findings) because large, diversified firms tend to have more products and processes to which new findings might be applicable (the second corollary). In this connection, Fisher and Temin note "this argument assumes there is no market for new knowledge *per se* . . . (which is not entirely true). . . . Pure knowledge, however, is not a widely traded commodity; it is linked to personnel, or machines or organizations." Third, the financial economies of scale enjoyed by larger firms lead to a greater willingness to take risks, partly because they have a lower risk of bankruptcy, lending support to the hypothesis and its first corollary.

Accordingly, we have included the first two corollaries with this hypothesis. This is in agreement with Kamien and Schwartz (1975, p. 8), who stated that the two significant scale effects to explore were the effect of firm size on R&D efficiency and the effect of the scale of R&D operations on R&D efficiency for a given size of firm. The third corollary is included because it reflects attempts by some economists to extend the posited relationship between the size of a firm and its level of R&D spending into studies of market structure.

The literature in these sectors of analysis makes frequent reference to "research intensity." Research input intensity is defined to be either the number of R&D personnel as a fraction of total employment or the proportion of sales revenue devoted to R&D expenditures. Research output intensity is defined to be the number of patents issued (sometimes *n* years later), or the number of significant inventions/innovations (sometimes *n* years later) divided by some measure of firm size (total assets, total employment, or sales). Studies of both research input intensity and research output intensity are used to estimate support for the hypothesis as it is reflected in the first corollary.

Research input intensity has been explored in several studies attempting to correlate it with the size of firm. Hamberg (1966) used employment variables to measure research intensity and both total employment and total assets as a measure of firm size in his study of 340 firms from the 1960 *Fortune* 500. He grouped these firms into 17 industries and found a positive correlation in only 3 industries. Comanor (1967) also used employment variables in his study of 387 firms in 21 industries with well-defined R&D programs, with findings even less definite than Hamberg's. Scherer (1965a) criticized such studies of R&D intensity for having samples biased toward large firms with large programs. He also indicated a preference for using sales rather than labor or total assets as a measure of firm size to avoid biases arising from variable factor proportions. Grabowski and Mueller (1970) went even further, judging the available data not to be high enough in quality to support a good test of research intensity versus firm size. In two recent surveys of the literature, Kennedy and Thirlwall (1972) concluded that no significant relationship between research intensity and size of firm was demonstrated by published studies, while Kamien and Schwartz (1975) were more generous in noting that the intensity of research effort increases only up to some level of firm size and then falls off, this critical size varying by industry and by study. Markham (1960) pointed out that the major problems with these studies include poor design, biased samples and missing variables; and Kamien and Schwartz (1975) reiterated these points in their review.

Similar efforts have been made to correlate the research output of firms (measured by patents or number of significant inventions) with firm size. Scherer (1965) used patents as a measure of research output, sales volume as a measure of firm size, and R&D employment as a measure of research effort for 352 firms from the 1955 *Fortune* 500. He found that research input intensity declined slightly for the largest firms relative to the smaller firms and that research productivity (patents divided by R&D employment) also declined slightly in moving from the smallest to the largest firms. Grabowski (1968) studied the R&D efforts of firms in chemicals, pharmaceuticals, and petroleum refining and found that a firm's research input intensity was positively correlated with its past research productivity (patents divided by R&D employment). Grabowski believed that removing the effect of past research productivity on research input intensity would be a good first step in removing the influence of interindustry differences in technological opportunities and also of interfirm differences in the capabilities of key R&D personnel. This assumes that past luck balances out across firms and industries, that top management groups would make similar decisions about R&D budgets when working with equal levels of knowledge about the effectiveness of R&D expenditures (as measured by past research productivity), and that they had equal levels of knowledge about this effectiveness at their disposal. But no evidence supports the widespread applicability of any of these assumptions.

Mansfield (1968) asked personnel in steel, chemicals, and coal mining to rank significant innovations developed from 1919 to 1950. Using productive capacity as a measure of output, Mansfield found research output intensity, measured by the number of significant innovations divided by productive capacity, to increase with relative firm size (actually, each industry was broken into two groups: the largest four firms and the rest of the industry) in chemicals and coal mining and to decrease with firm size in steel. Freeman (1974) studied more than 1,200 post-World War II innovations in Britain; he found that firms with less than 200 employees contributed about 10 percent of the total innovations while accounting for 21 percent of net output and 25 percent of total employment. More interestingly, he found this result was reversed in industries where the cost of entry and capital intensity were low.

Kamien and Schwartz (1975, p. 19) believe that these studies have shown a "critical threshold" level of firm size beyond which innovational output rises less rapidly than firm size, although this critical level varies by industry. This is in accordance with their summary of the relationship between firm size and research *input* intensity and reflects the meager findings connecting research input with research output noted in the previous section.

The most common approach of studies attempting to include the effect of market structure is to use industry concentration—measured by the fraction of total industry sales accounted for by the top four firms (see, for example, Hamberg, 1963, and Scherer, 1965b), or the top eight firms (see Comanor, 1967) in the industry—as an explanatory variable for the level of research (input or output) intensity found in a given industry. For example, Comanor (1967) attempted to estimate the effect of both absolute average firm size (measured by sales) and "eight-firm market concentration" ratios in his study of 387 firms in twenty-one industries. Concentration was found not to be significant in explaining industry differences in research intensity (as measured by employment), whereas average firm size was found to be significantly correlated, indicating support for the first corollary but no support for the third. Comanor believed it was possible that technological opportunities varied across his twenty-one industry sample. Associating technological opportunities with the ease of achieving product differentiation, he used the McGraw-Hill surveys on the purposes of R&D to support his contention that a major goal of R&D is to develop new products to achieve a protected market position—that is, to achieve greater product differentiation. He segregated his twenty-one industry sample into a group with high potential for product differentiation, consumer and producer durables, and a group deemed to have little such potential, consumer nondurable goods and materials. These same industries were also segregated into two groups on the basis of the eight-firm concentration ratio (above 70 percent and below 70 percent). Research intensity in each firm was adjusted for the predicted research intensity in each industry, based on the first part of the study reviewed above. Employing his 2 X 2 classification on opportunities and concentration, Comanor concluded that (a) research intensity was much higher on average in industries with high technological opportunities; and (b) that higher research intensity was associated with higher concentration only in industries with low technological opportunities. It is difficult to understand, however, how Comanor identified consumer and producer durables in general as industries in which opportunities for product differentiation are high and consumer nondurables as industries possessing only limited potential in this respect. The connection between *technological* opportunity and product differentiation is also not clear.

Scherer (1967) used employment variables to measure firm size and R&D input in his study of concentration. In one approach, he attempted to adjust for technological opportunities on the primary basis of interindustry differences in patents granted in 1959 that were not explained by interindustry differences in sales in 1955. He recognized four classes of technological opportunity on the basis of this technique. In his subsequent analysis, he regressed an index of R&D employment against concentration, total industry employment, and a number of qualitative variables that defined the type of goods (producer, consumer, durable, nondurable) and the technological opportunity class. Concentration ranged from very significant, when the qualitative variables were omitted, to moderately significant when they were included. Concentration remained moderately significant when he used R&D input intensity as his dependent variable and adjusted for the effect of technological opportunities. Scherer concluded that (a) concentration was significantly and positively associated with R&D intensity; and (b) the positive correlation between technological opportunity and concentration was (1) responsible for the reduction in the significance of concentration in his first regression following the inclusion of "technological opportunity class" as an additional independent variable; and (2) indicative that a greater pace of technological innovation may have resulted in increased concentration in industries with high technological opportunities. Of course, causality is not unambiguously unidirectional. If most firms in a particular industry persisted in believing they enjoyed high technological opportunities, the effect might be the same.

Despite the moderately positive support for the third corollary derived from these studies, Kennedy and Thirlwall (1972) and Kamien and Schwartz (1975) concluded on the basis of much wider surveys that the evidence in this area is inconclusive. Given the weaknesses of data, coverage, theory, and models, we agree with these conclusions; we also wish to emphasize that research in this area so far represents only fragmentary explorations in seeking to answer a variety of enormously difficult and complex questions.

At present, we do not have a satisfactory theory that would explain why larger firms normally spend a greater fraction of their sales revenues on R&D than smaller firms. Consequently, because a number of potential explanations exist, it would certainly be helpful if future empirical research investigating research intensity were directly focused on several disparate possibilities.

1. *Diversification:* The extent to which larger firms find higher relative or absolute R&D budgets more attractive because the uncertain findings of R&D programs have a greater probability of payoff when the firms' products and processes are more diversified.

2. *Scale:* The share of total production that represents product and process areas in which firms believe that there are economies of scale in R&D activities. Some evidence would have to be uncovered that indicates that for larger firms, marginal research benefits approach marginal research costs (assuming these concepts are measurable and sensible at all) only after the larger firms spend proportionately larger amounts of sales revenues on R&D. In this case, one might question why small firms would bother to spend anything on R&D. Possible answers are that marginal research benefits grow more than proportionately with firm size, that larger firms have greater technical problems to overcome because their plants are designed at a higher scale of operations, and that small firms do not focus on the same spectrum of R&D problems as do large firms.

3. *A combination of diversification and scale:* If there is a critical threshold level of R&D expenditures in specific product or process areas such that lower levels of expenditures have unattractive payoffs and higher levels of spending begin to yield handsome expected profits, and if this level varies by type of product or process (including the scale of production) then higher levels of research intensity could be associated with larger, more diversified firms; up to some arbitrary level of firm size.

4. A logic that is the reverse of the above three arguments could also be important: namely, that larger firms tend on average to enjoy higher profit margins and that R&D activities are budgeted on the basis of profits earned. In this view, the profitability of R&D activities is regarded as highly uncertain and speculative, hence R&D is treated as a "luxury" activity.

The literature concerned with the third hypothesis has also raised some interesting questions concerning the effect of market structure and technological opportunities on the link between firm size and research intensity. Because the definition of a market is certainly subject to some degree of arbitrariness, and because current definitions seem more concerned with product substitutability than with process similarities, it seems unreasonable to expect to find clearcut evidence on the nature of these effects. More specifically, the use of concentration ratios (supplemented, perhaps, with qualitative variables relating to basic product or process characteristics) as a measure of market structure seems unlikely to yield anything more than oblique illumination of the connection between firm size and research intensity. At the very least it would be helpful to know how current patterns of market concentration arose.

Classifying the technological opportunities of a market (industry) on the basis of past innovational successes is clearly not an objective method of adjusting for such opportunities, unless one subscribes to the "mechanistic" school of thought on the origins of innovations. Yet, because the history of science and technology is not a repeatable experiment, it is difficult to see how current measures of technological opportunity classes can be improved. Consequently, future research efforts incorporating this elusive concept must be taken with the proverbial grain of salt.

Notes

1. Among the materials we have found particularly valuable should be included: the National Bureau of Economic Research Conference volume, edited by R.R. Nelson (1962); the Proceedings of the Ohio State University Conference on the Economics of Research and Development, edited by R. Tybout (1965); the wide-ranging studies of Carter and Williams (1957, 1958) and Jewkes, Sawers, and Stillerman (1969); the attempts at quantitative evaluations by Mansfield (1965); and recent survey articles by Kennedy and Thirlwall (1972) and Kamien and Schwartz (1975).

2. Aircraft and parts, electrical machinery, vehicles, chemicals, fabrication of metals, stone, clay and glass, textiles, food and beverages, petroleum and coal products, nonferrous metals, iron and steel, and rubber.

3. For an increase in total factor productivity to be reflected in increased average profits, the price of products must remain constant, relative to the prices of inputs. Minasian (1962) offered no assurances that this condition was fulfilled in his study.

4. That is by L^aK^{1-a} where a is measured by the same method outlined earlier in this chapter, where L is labor and K is capital. For a review of the specific measures of capital, labor, and output (that is, value-added), see earlier sections.

5. The longer lag periods could only be used when total factor productivity and relative R&D expenditures were measured over the shorter intervals.

6. See note 3.

7. Defined at the two-digit SIC level of aggregation: chemicals, machinery, food, paper, instruments, electrical equipment, stone, clay, and glass, apparel, motor vehicles, and furniture.

References

Ames, E., "Research, Invention, Development and Innovation," *American Economic Review* (June 1961).

Arrow, J.K., "Classificatory Notes on the Production and Transmission of Technological Knowledge," *American Economic Review* (May 1969).

Brozen, Y., "R&D Differences Among Industries," in *Economics of Research and Development*, ed. R.A. Tybout (Columbus: Ohio University Press, 1965).

Carter, C.F., and B.R. Williams, *Industry and Technical Progress* (London: Oxford University Press, 1957).

_______, *Investment in Innovation* (Oxford University Press, 1958).

Comanor, W.S., "Research and Technical Change in the Pharmaceutical Industry," *Review of Economics and Statistics* (May 1965).

_______, "Market Structure, Product Differentiation and Industrial Research," *Quarterly Journal of Economics* (November 1967).

_______, and F.M. Scherer, "Patent Statistics as a Measure of Technical Change," *Journal of Political Economy* (May/June 1969).

Enos, J.L., "Invention and Innovation in the Petroleum Refining Industry," in *The Rate and Direction of Inventive Activity: Economic and Social Factors*, ed. R.R. Nelson (Princeton, N.J.: Princeton University Press, 1962).

Fisher, F.M., and P. Temin, "Returns to Scale in R&D: What does the Schumpeterian Hypothesis Imply?" *Journal of Political Economy* (January/February 1973).

Freeman, C., *The Economics of Industrial Innovation* (Baltimore: Penguin Books, 1974).

Grabowski, H.G., "The Determinants of Industrial Research and Development: A Study of the Chemical, Drug, and Petroleum Industries," *Journal of Political Economy* (March/April 1968).

Grabowski, H.G., and K. Mueller, "Industrial Organization: The Role and Contribution of Econometrics," *American Economic Review, Papers and Proceedings* (May 1970).

Griliches, Z., "Research Costs and Social Returns: Hybrid Corn and Related Innovations," *Journal of Political Economy* (October 1958).

_______, "Research Expenditures, Education and Aggregate Agricultural Production Function," *American Economic Review* (December 1964).

Hamberg, D., "Invention in the Industrial Research Laboratory," *Journal of Political Economy* (April 1963).

_______, *R&D: Essays on the Economics of Research and Development* (New York: Random House, 1966).

Hirsch, W.Z., "Technological Progress and Microeconomic Theory," *American Economic Review, Papers and Proceedings* (May 1969).

Hitch, C.J., "Comment," in *The Rate and Direction of Inventive Activity* (1962).

Jewkes, J., D. Sawers, and R. Stillerman, *The Sources of Invention*, 2nd ed. (New York: Norton, 1969).

Kamien, M.I., and N.L. Schwartz, "Market Structure and Innovation," *Journal of Economic Literature* (March 1975).

Kennedy, C., and A.P. Thirlwall, "Surveys in Applied Economics: Technical Progress," *Economic Journal* (March 1972).

Kuznets, S., "Inventive Activity: Problems of Definition and Measurement," in *The Rate and Direction of Inventive Activity* (1962).

Lynn, F., "An Investigation of the Rate of Development and Diffusion of Technology in Our Modern Society," in *Studies Prepared for the National Commission on Technology, Automation and Economic Progress*, Appendix, Vol. 2 (Washington, D.C.: U.S. Government Printing Office, 1966).

Maclaurin, W.R., "Technological Progress in Some American Industries," *American Economic Review* (May 1954).

Machlup, F., "The Supply of Inventors and Inventions," in *The Rate and Direction of Inventive Activity* (1962).

Mansfield, E., "Rates of Return from Industrial Research and Development," *American Economic Review, Papers and Proceedings* (May 1965).

_______, *The Economics of Technological Change* (New York: Norton, 1968).

Markham, J., "Economic Analysis and the Research and Development Decision," in *Economics of Research and Development* (1965).

Minasian, J., "The Economics of Research and Development," in *The Rate and Direction of Inventive Activity* (1962).

_______, "Research and Development Production Functions and Rates of Return," *American Economic Review*, May 1969.

Mueller, W.F., "The Origins of the Basic Inventions Underlying DuPont's Major Product and Process Innovations, 1920-1950," in *The Rate and Direction of Inventive Activity* (1962).

Nelson, R.R. (ed.), *The Rate and Direction of Inventive Activity: Economic and Social Factors* (Princeton, N.J.: Princeton University Press, 1962)

National Science Foundation, *Review of Data on Science Resources* (various issues).

Pavitt, K., and S. Wald, *The Conditions for Success in Technological Innovation* (Paris: OECD, 1971).

Peck, M.J., "Inventions in the Postwar American Aluminum Industry," in *The Rate and Direction of Inventive Activity* (1962).

Sanders, B.S., "Some Difficulties in Measuring Inventive Activity," in *The Rate and Direction of Inventive Activity* (1962).

Sanow, K., "Development of Statistics Relating to Research and Development Activities in Private Industry," in *Methodology of Statistics on Research and Development* (Washington, D.C.: National Science Foundation, 1959).

Scherer, F.M., "Size of Firm, Oligopoly, and Research: A Comment," *Canadian Journal of Economics and Political Science* (May 1965a).

_______, "Firm Size, Market Structure, Opportunity and the Output of Patented Inventions," *American Economic Review* (December 1965b).

_______, "Market Structure and the Employment of Scientists and Engineers," *American Economic Review* (June 1967).

Schmookler, J., "The Changing Efficiency of the American Economy, 1869-1938," *Review of Economics and Statistics* (August 1952).

_______, "The Level of Inventive Activity," *Review of Economics and Statistics* (May 1954).

_______, *Invention and Economic Growth* (Cambridge, Mass.: Harvard University Press, 1966).

Silk, L., *The Research Revolution* (New York: McGraw-Hill, 1960).

Tybout, R.A. (ed.), *Economics of Research and Development* (Columbus, Ohio: Ohio State University Press, 1965).

Usher, A.P., *A History of Mechanical Inventions* (Cambridge: Harvard University Press, 1954).

6 Reported Economic Effects of Technological Changes

Myles G. Boylan

This review of the economic effects of technological changes covers the most important "modern" theoretical contributions since the early 1930s as well as empirical studies since 1950. Its initial focus was on all direct economic effects of specific innovations, but such studies were found to be very scarce. Consequently, coverage was broadened to include samples of studies investigating the effects on one or more elements of inputs and costs of a stream of innovations adopted over definite periods of time. Among these, special attention has been given to the large cluster of studies concerned with the effects of specific innovations on the displacement of labor.

The first section surveys the theory of the effects of process innovations. The following sections examine the findings of empirical studies of a stream of innovations over some period of time, the labor input, single innovations, and the diffusion of process and product innovations. The final section presents our conclusions.

Basic Concepts and Measures of Technological Change

Technological Change Defined

Technological change may be defined directly, or it can be defined indirectly in terms of its effects on the productivities of inputs.[1] Mansfield (1968, pp. 10-11) has offered the following direct definition:

> Technological change is the advance of technology,[2] such advance often taking the form of new methods of producing existing products, new designs which enable the production of products with important new characteristics, and new techniques of organization, marketing and management.

Schumpeter (1939, pp. 87-88) defined technological change as synonymous with innovations; he also defined innovations indirectly in terms of their effects on input requirements:

Joel Rosenberg, Assistant Professor of Economics, participated in the early review of the literature and William Peirce, associate professor of economics, contributed pp. 144-46.

> We will now define innovation more rigorously by means of the production function—this function describes the way in which quantity of products varies if quantities of factors vary. If, instead of quantities of factors, we vary the form of the function, we have an innovation.

Solow (1957, p. 312) employed essentially the same indirect definition of technological change in his pioneering article: "I am using the phrase 'technical change' as a shorthand expression *for any kind of a shift* in the production function."

Schumpeter (1939, p. 84), Ruttan (1959, p. 605), and Mansfield (1966, p. 11) agree that many innovations (that is, technological changes) are not derived directly from predecessor inventions and, in fact, that they occur with a high frequency greatly exceeding that of well-publicized, major innovations that *do* depend to some degree on predecessor inventions. Indeed, Ruttan (1959, p. 606) adopted an extreme position by suggesting that it was pointless to separate analytically inventions from innovations.[3] Instead, he believed that inventions should be viewed as a subset of innovations and that the latter could be defined both directly as "the entire range of processes . . . by which 'new things' emerge in science, technology and art"; and indirectly "technological change . . . designates changes in the coefficients of a function relating inputs to outputs resulting from the practical application of innovations in technology and in economic organization."

Expected Effects of Technological Changes

Blaug (1963, p. 13) believed it was useful to categorize technological change into two classes—product and process innovations—although he recognized the distinction was largely artificial. He noted that practically all theoretical treatments of technological change were theories of process innovations, because:

> Since the index-number problem has so far doomed all theoretical analysis of innovations which alter the quality of final output, the refusal to discriminate between product—and process—innovations would close the subject of technical progress to further analysis.

This statement has proven to be largely correct.[4]

Many studies since the 1930s have attempted to predict and classify the effects of a process innovation. Hicks (1936), Salter (1960), Mansfield (1966), and others have employed the concept of unit isoquants to illustrate the effects on unit input requirements. If we abstractly define the production inputs as belonging to one of two categories—capital or labor—the unit isoquant reflects the various combinations of capital/output and labor/output that can be employed to produce a given amount of output.[5] In conventional analysis, it is universally assumed that the initial effect of a process innovation is to shift the unit isoquant closer to the origin. In Figure 6-1, this is indicated as a shift from I_0 to I_1. Firms adopting the innovation might find that the average productivity of both capital and labor (defined as output/capital and output/labor) have increased, but they will *always* find that at least one of these has increased, because the initial expected effect of the innovation is a reduction in total unit cost because factor prices are assumed to be the same before and after the innovation has been adopted. In fact, the anticipated reduction in total unit cost is assumed to be the basic motivation for adopting the innovation.

A large body of literature attempts to go beyond these seemingly obvious effects and to categorize (and even predict) the new "equilibrium position" of the firm after the diffusion process is complete. For example, if firms were initially in equilibrium at point *A* on the unit isoquant labeled I_0, they would end up at point *B* on I_1 after the innovation is completely diffused, assuming no effect on factor prices.[6] (The slope of I_1 at *B* is equal to the slope of I_0 at *A*). But, as Hicks (1936), Gold (1964), Blaug (1963), and others have noted, this is very unlikely. According to Hicks, if a stream of innovations saved labor relative to capital in the total economy, the effect would be a reduction in the average wage rate relative to the "price" of capital (such as the prime interest rate), which would in turn induce firms to substitute labor for capital moving them from point *B* to point *C*. Gold has noted that if trade unions have become accustomed to receiving increased wages when output/labor is increased, the

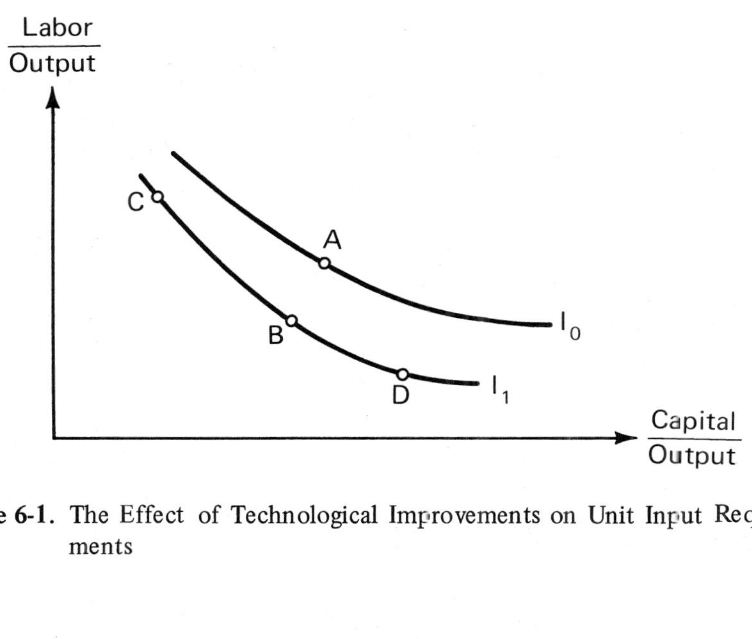

Figure 6-1. The Effect of Technological Improvements on Unit Input Requirements

opposite effect would occur within an industry, and firms would be induced to move from point B to D.[7]

Blaug reminds us that the innovation is ultimately likely to result in a relative reduction in the price of the affected product, thereby increasing the quantity demanded and sold. Unless constant returns to scale prevail after the innovation has diffused, this will change the capital to labor ratio and, hence, move us from point B on I_1 to another point on another unit isoquant. Gold has gone further in pointing out that it is extremely difficult to predict the ultimate effect of a given innovation on sales, because other innovations of parallel importance in the same or related industries shift the demand curves both for the product in question and for the factors used in producing them.

In short, although it is generally agreed that factor prices and the product's price will be affected by a process innovation, there is great uncertainty about the relative magnitude and direction of these price effects. We believe that this uncertainty is in accordance with the great diversity among innovations and among the industries and markets affected by them. The important consequence of associated factor and product price adjustments is that initial changes in unit input requirements give only temporary clues about the magnitude of eventually realized changes in costs and profits.

Measures of the Effects of Technological Changes

The commonly used measures of the effects of technological change concentrate on productivity changes defined in physical terms. According to Mansfield (1966), for example, three commonly used measures include (a) the growth in labor productivity; (b) growth in total factor productivity; and (c) the ratio of best-practice to average-practice productivity. As we explained in Chapter 2, growth in total factor productivity can be related to shifts in production functions and, thus, is akin to measuring the residual growth in output not explained by increases in the volume of inputs (see, for example, Kennedy and Thirlwall, 1972; Brown, 1968). Enos (1962) used the shifts in the average productivities of capital, labor, energy, and materials to illustrate the effects of several important twentieth-century process innovations in petroleum refining. Salter (1960) suggested using the ratio of best-practice to average-practice productivity to illustrate the potential gains when a typical firm adopted a new process. But most studies do not even purport to describe the full range of effects of technological change and concentrate only on selected aspects. For example, many studies have emphasized the effects of technological innovations on labor, one group concentrating on resulting changes in average labor productivity (see, for example, Weinberg, 1962; Johnson, 1972; and Butcher, 1966) and another on the displacement of labor by new automated processes.

Measuring the physical effects of process innovations represents a necessary and important first step in describing the total effects of such innovations. Even this task, however, is not as simple as is often supposed. Mansfield (1966) criticized total factor productivity measures because they attribute all residual improvements (gains) to technical change when such change could also be the result of changes in product mix, improvements in labor skills and health, economies of scale, or improved resource allocation. He also noted that attempts to measure all productivity changes necessitate that capital productivity be included, thus supporting Gold's earlier position (1955, pp. 59-80, 135-145). But capital is very difficult to measure (Gold, 1955, pp. 18-30), and simple measurements are inadequate even in respect to labor, to say nothing of the frequently complex array of material inputs. In addition, Gold (1973) has emphasized that it is often not possible to measure the physical effects of separate innovations at the industry or even the plant level of aggregation because of the tendency of industries to adopt a number of overlapping innovations during the extended time periods involved in fully adjusting to the effects of major innovations. Finally, if the initial effects of an innovation cause changes in factor prices, and firms subsequently readjust their input mixes to such changes in relative factor prices, it is difficult to determine what fractions (if any) of such readjustments are reflected in the physical measures of distinctively technological effects.

Another very common measure of the effects of technical change is to estimate the reduction in total unit cost (and its components). Both Mansfield (1966) and Salter (1960) have stressed that this measure is of primary interest, both *ex ante* and *ex post*, to the potential and actual adopters of process innovations. They believe that, although changes in input-output ratios may be of some interest to the firm, the reduction in total unit cost is of paramount importance to the firm. Examples of this approach can be found in Enos (1962), Maddala and Knight (1967), Dilley and McBride (1967), and Hollander (1965). Mansfield (1966) also reasons that although firms understand that reductions in those factors that account for the largest shares of total cost will have the greatest impact on reducing total unit cost, the typical firm cannot predict very precisely what kind of factor-saving bias will result from its efforts to improve technology.

One weakness of relying on changes in total unit costs is that such costs are responsive to many sources of continuing change in addition to the initial innovational impacts. These include exogenous changes in factor qualities, supplies, and prices, as well as in the level and composition of product demand and also in competitive pressures. Such multidimensional interactions obviously make it difficult to isolate the changes in such costs attributable to the distinctive effects of the innovation being studied. Accordingly, many studies have retreated to attempting to measure the initial decline in total unit cost recorded by early adopters of the innovation. This is regarded as providing at least a rough picture of the benefits of the innovation and the resulting competitive pressure on other firms to adopt the innovations.[8]

In one of the more complex approaches to measuring the effect of a process innovation (or a stream of innovations) on the net rates of change in unit labor and unit capital requirements, Salter (1960) demonstrated that these net rates of change were comprised of three separate effects: (a) the percentage reduction in total unit cost, assuming no change in factor prices and no bias in the rate of factor savings (which is equivalent to the inverse of the increase in total factor productivity); (b) a bias effect, which is defined as a change in the capital to labor ratio due *solely* to the innovation (that is, *not* caused by a change in relative factor prices); and (c) a substitution effect brought about by changed factor prices and a change in the degree of ease with which capital and labor can be interchanged.[9] However, as Brown (1968) has noted, Salter's measurement model is an analytical curiosity rather than a concrete aid to this measurement problem because its last two components cannot be measured independently from the first component. Indeed, Salter himself abandoned his framework in later empirical work.

In summary, the measures that have been employed have included changes in (a) the average productivities of specific inputs; (b) total factor productivity; and (c) unit costs, particularly total unit cost. No studies were found that were successful in isolating the effects generated by the given innovation from subsequent movements in relative factor prices and product prices, and further readjustments in adjacent as well as more distant operations.

Empirical Studies of the Effects of All Technological Changes During a Period

The empirical literature on the effects of technological innovations has been much more frequently concerned with resulting changes in the average productivity of inputs than in factor prices and unit costs. Although this suggests that it has been easier to obtain data on physical than on financial flows, many studies have actually employed time series data on the value of materials and capital employed and deflated them by some price series to approximate the "physical" flows of these two inputs—thus mixing together the problems of measuring productivity changes with those involved in measuring changes in the prices of complex arrays of materials and of capital facilities and equipment.

Within this literature, the primary focus is on the combined effects of a stream of innovations over a time period rather than on the effects of particular major innovations. At one extreme, similar to studies of the aggregate production function for all manufacturing, are the well-known pioneering contributions of Fabricant (1940) and Kendrick (1961), whose data have been widely employed in the studies of the aggregate production function reviewed in Chapter 2. At a less aggregative level of analysis are such studies of samples of plants, as Weinberg's analysis of three pulp and paper plants in the 1950s (1962) and Hollander's study of five DuPont rayon plants (1965). Relevant publications also include a few analyses of the effects of specific innovations; most of these occur in the steel and petroleum refining industries.

Industry-Level Studies

Both Butcher (1966) and Bauer (1964) attempted to assess the effects of technological change on employment in agriculture. Butcher's research encompassed the entire agricultural sector in the 1890-1970 period, whereas Bauer limited his study to thirty-nine states during the 1951-1961 period. Butcher measured total factor productivity by Kendrick's arithmetic formula, as reviewed in Chapter 2: labor in terms of man-hours, other inputs (machinery, motor supplies, and fertilizers), as well as output in terms of deflated values. Within the limitations of these data and measures, Butcher showed that output per man-hour rose 5 percent annually between 1940 and 1970, while total factor productivity rose more than 2 percent annually. Hence, according to Butcher, other inputs were substituted for labor; the implication is that the growth in labor productivity alone overstates the effect of technological change. He did not attempt to link these two productivity measures and left unanswered such questions as whether the shift from labor to other inputs was induced by the stream of innovations, by shifts in relative factor prices, or by still other influences.

Bauer employed a relatively sophisticated simultaneous equation model of the supply and demand for farm labor to assess the effects of technology. The demand for farm labor was defined as responsive to the farm wage rate, the deflated value of nonlabor expenditures in agriculture, and "technology." The latter was measured by state and federal expenditures on experiment station research and the Cooperative Extension Service; this figure obviously has only a tenuous relationship to technological changes or even to R&D expenditures directed toward such ends. At any rate, this measure of technology was expected to influence the demand for labor over several subsequent years; its major impact was estimated to have occurred one year later. Because Bauer assumed, in the same manner as Griliches (1958), that the initial effect of technology was to increase farm output—given the stock of land, the size of the labor force (and the wage rate), and the level of other inputs—the ultimate impact on the demand for labor depended on the elasticity of demand for farm output. If the entire increase in farm output was absorbed at unchanged farm prices, implying that the demand for output was completely elastic, Bauer estimated that the derived demand for labor would rise 1 percent for each 3 percent increase in government research and extension service expenditures.[10] If, as is more likely, the demand for farm output was inelastic, then increases in government research and extension service expenditures would have a negative impact on employment or wage rates or both.

Bauer's findings seem to rest on very shaky conceptual foundations. The weakest feature of his model is his assumption that the flow of innovations (lagged from one year to decades later) matches the flow of government expenditures for experiment stations and for the wide-ranging Extension Service—a view whose unlikelihood was discussed in Chapter 5. In addition, by not specifying the impact of innovations on other inputs in agriculture, his model apparently assumed them to be unaffected, only changing exogenously over time—even though it is far more likely that a number of innovations affected labor only indirectly through changing the qualities of and methods of using other inputs. The most notable feature of his model was the inclusion of the factor price of labor, but this was not directly linked to labor productivity.

In the railroad transportation sector of the United States economy, Johnson (1972) and Peterson (1968) studied the impact of such innovations as diesel locomotives, freight car improvements, and improved controls, the latter covering an expanding use of computers, automatic classification yards and central traffic controls. Johnson's study covered all railroad services in the 1947-1968 period, whereas Peterson's investigation was limited to Class I line-haul railroads that accounted for roughly 90 percent of all line-haul revenues in the shorter 1947-1963 period. Because the major post-World War II innovations had widely diffused by 1963, and because both studies employed almost the same methodology, these studies can reasonably be compared.

Johnson's approach to capturing the effects of these innovations was based primarily on measuring changes in the productivities of capital, labor, and materials (including fuel) with some links to factor prices. Output was measured by a weighted average of revenue ton-miles (for freight) and revenue passenger-miles. Capital was measured as the net value of plant and equipment in 1958 dollars. Purchased materials, covering rails, ties, and fuel, were measured in 1958 dollars and labor was measured in man-hours. Peterson measured the productivities *and costs* of labor, capital, and materials (including fuel); in some cases, he further disaggregated these measures of performance. They were measured similarly to Johnson's methods. Output, however, was measured by weighting the quantity of each service class (for both passenger and freight service), by average prices for each class of service. The resulting index of output is, therefore, adjusted for changes in service mix and is superior to Johnson's estimates.[11] Accordingly, differences in the findings of these two studies can be attributed to the different measures of output employed and the five additional years covered by Johnson.

Peterson found that labor productivity had doubled (Johnson: tripled), materials productivity increased by 50 percent (Johnson: tripled), and capital productivity had declined progressively to 70 percent of its 1947 level (Johnson found no change in capital productivity). Because it is doubtful that the 1963-1968 period could account for the much higher net productivity changes reported by Johnson, it may be surmised that a shift in product mix toward lower value service classes was not captured by Johnson's measure of output (that is, his measure progressively overstated output). Petersen attributed the decline in capital productivity to the railroads' strategy of modernizing equipment without any attendant increases in total capacity. Johnson claimed that part of the increased labor productivity was due to effects embodied in the scale and technology of capital equipment; for example, more powerful diesel locomotives and improved rail car design resulted in 18 percent heavier average freight car loads and 24 percent longer trains, resulting in a 50 percent gain in average train size without a comparable increase in crew size.

By incorporating factor prices in his study, Petersen was able to demonstrate that Johnson's finding, which suggests that some relative factor savings may be caused by "biased" technological change, probably characterized most of the innovations underlying changes in input productivities reported previously. He estimated that capital equipment prices rose 75 percent, materials (including fuel) prices rose 75 percent; and average wages and salaries rose 125 percent. These changes resulted in approximately 15 percent gains in unit labor and unit materials costs, whereas unit overhead cost rose by more than 100 percent. This apparent excess substitution of capital for labor can be reasonably explained, however, on a less aggregate level of analysis. For example, unit fuel cost alone decreased by 40 percent according to Petersen, primarily as a result of the diffusion and improvements in diesel locomotives and despite a 25 percent increase in the price of diesel oil. Unit track maintenance costs did not rise nearly as rapidly as they would have if railroad management had not allowed the physical condition of many sections of track to deteriorate. Indeed, it may be surmised that this policy, involving a reduction in labor-intensive activities, also improved apparent labor productivity. Similarly, improved controls of various types led to apparent declines in capital productivity and measured improvements in labor productivity. In short, capital-labor (and capital-materials) substitution was initiated by capital-capital substitution because new technology tended to be embodied in new facilities and equipment. A major question, not raised in either study, concerns the relative effects on the rate of introduction of available innovations of access to needed capital; the magnitude of expected savings in unit operating costs; labor resistance to attendant changes; and prospective improvements in customer service and in system utilization.

Maddala studied productivity and technical change in the bituminous coal mining industry from 1919 to 1954 (1965). He estimated capital and labor productivity following a procedure for utilizing Census of Manufactures data on the horsepower of installed equipment suggested earlier by Phillips (1956) and used even earlier by Jerome (1934). Phillips had suggested that horsepower per unit of product and horsepower per wage-earner could be used as an indicator of technological change (under the assumption that horsepower was a proxy for capital) if used carefully in conjunction with output per man-hour. Maddala concluded that the rise in output per man-hour could be explained almost completely by the increase in horsepower per worker (that is, by capital-labor substitution) when he connected these two inputs with output in a Cobb-Douglas production function. Thus, he ignored all the qualitative dimensions of technological improvements as well as strong evidence that such improvements had been made, and failed to heed Phillips's warnings of the need to use horsepower data cautiously. Furthermore, without a price series for capital, it is difficult to see how he can claim the perceived capital-labor substitution was not a result induced by coal mining innovations only partially reflected by horsepower increases. Finally, he neglected the possibility that the bituminous coal reserves may have become increasingly difficult to mine, thus offsetting some of the productivity gains that would have occurred if coal deposits had been as accessible in 1954 as they were in 1919.

Gold (1964) measured unit input cost, input cost proportions and unit input requirements for six industries during the 1899-1939 period with the primary objective of investigating possible connections between differences in productivity improvements and changes in relative factor prices for major classes of inputs. First, he employed data from the Census of Manufactures to show that the proportions of total cost accounted for by materials, hourly labor, salaried employees, and capital in each of six industries (blast furnace products, steel mill products, coke oven products, cement, petroleum refining, and canned fruits and vegetables) were remarkably stable during the forty years covered. This finding indicated a good possibility that there were consistent interconnections between unit input requirements and the prices of the inputs.

Accordingly, Gold compared changes in the unit labor and unit materials requirements with changes in their respective prices at (approximately) five-year intervals from 1909 to 1939 for three industries (coke oven products, blast furnace products, and petroleum refining). The frequency of reductions in unit labor requirements was 75 percent, and unit material requirements fell 95 percent of the time, yielding significant evidence of technological progress. Gold offered initial hypotheses that: (a) improved labor productivity would be offset by wage gains resulting from piece rates and the pressures of actual or prospective trade unions; and (b) increased productivity of materials would result in reduced materials prices in *competitive* materials markets for comparable levels of output; but (3) changes in capital productivity would have little *secondary* effects on unit capital costs because the market for capital is economy-wide. The finding of stable cost proportions clearly supports the initial labor-wage hypothesis but contradicts the hypothesis that materials-saving innovations would also lower materials prices.[12] These symmetrical, but partially contradictory findings are hardly surprising, according to Gold, because although the prevailing hypotheses emphasize uni-directional causation from changes in input productivities to changes in factor prices, it seems obvious that a rounded hypothesis would also encompass the effects of changes in factor prices on unit input requirements.

A more detailed investigation of the interaction between changed input productivities and changed input prices revealed that increased labor productivity engendered higher wage rates more frequently and more rapidly than the latter induced increases in labor productivity. In the case of materials, however, price increases induced reductions in unit input requirements much more frequently than the latter engendered lower materials prices. During further analysis and discussions, Gold (1964) argues that the interconnections between unit input requirements and input prices are far more complex than indicated by these foregoing hypotheses; that it may not be possible to generalize across all manufacturing industries; and that perceived interconnections depend on (and influence) the rate of diffusion of the innovation as well as the time periods over which changes are measured.

Industry and Plant-Level Studies

Zeisel (1968) investigated the effects of a stream of innovations in the textile industry from the mid-1940s to the mid-1960s. She measured these effects at the plant and industry levels by comparing movements in labor productivity,[13] capital productivity,[14] and energy productivity[15] with corresponding changes in the capital-labor ratio (horsepower per employee), net fixed investment, and "efficiency" (defined as a decrease in the ratio of payroll to value added).

Unfortunately, Zeisel never demonstrated that reductions in the ratio of payroll to value added were highly correlated with relative reductions in total unit cost (which is the standard measure of efficiency) at the plant level. Obviously, however, reductions in this ratio could be associated with increases in total unit costs under a variety of common conditions, including reductions in the utilization of plant capacity; major increases in depreciation and interest charges because of investments in new facilities and equipment; or because of increases in the cost of materials resulting from higher quality or more stringent supplies. Thus her apparent assumption can hardly be accepted as generally applicable that most technical improvements are *embodied* in new equipment, and that resulting reductions in unit labor requirements generate reductions in unit labor costs greater than accompanying increases in net fixed investment charges.

At the industry level, Zeisel found that energy productivity was practically unchanged, capital productivity rose by roughly 100 percent according to both measures employed, and labor productivity rose by more than 100 percent (in some plants, by 200 percent). Although her finding that increases in labor productivity exceeded increases in capital productivity parallels similar findings by Butcher (1966) for agriculture and by Petersen (1968) and Johnson (1972) for railroads, the increase measured in the capital-labor ratio was much smaller.

At the plant level, Zeisel found that those plants with the largest increases in net fixed investment also experienced the largest increases in capital-labor ratios, resulting in the highest increases in labor productivity, practically no increases in capital productivity, and major increases in "efficiency." Practically all these correlations are meaningless. However, an estimate of the change in the labor and capital cost proportions would have confirmed or denied the value of her definition of efficiency. For example, in a study of productivity trends in the basic steel industry between 1904 and 1969, Gold (1974) showed, over four subperiods, that changes in capital productivity were inversely correlated with changes in the capital-labor ratio, and the variance of these two variables was very large, while advances in labor productivity were much steadier.[16] He then demonstrated that in the subperiods in which the capital-labor ratio increased the most—the same subperiods in which capital productivity barely advanced or even fell—that total unit cost was more likely to rise, and increases in capital charges were likely to exceed proportionately any increases in total wage costs brought about by increased wage rates. This finding suggests that Zeisel's definition of efficiency may have been just the reverse of what it should have been.

Plant-Level Studies

Weinberg (1962) investigated the effects of three classes of innovations in three pulp and paper mills during the 1950s. These classes of innovations were (a) mechanization of materials handling, (b) continuous processing equipment, and (c) automatic paper roll handling systems. He found these innovations required relatively heavy increases in investment, leading to a decline in the productivity of capital (measured by the ratio of physical output to capital, where capital was measured by current net fixed investment or in horsepower). Labor productivity (measured by physical output per man-hour) increased by 25 percent from 1954 to 1958. After adjusting wage rates for inflation during the 1954-1956 period by a technique only vaguely described, Weinberg found the increase in adjusted wage rates (which supposedly reflected shifts to higher skill classes of labor) to be well below the gains in output per man-hour. However, his assumption that all wage adjustments for gains in labor productivity would have taken place immediately in the 1954-1956 period is questionable; also, whether any of the actual increases in wage rates was eliminated during his deflation adjustment is not specified.

Hollander (1965) studied the effects of innovations in five DuPont viscose rayon plants over various periods within the major 1929-1960 period. He used a 2 × 2 classification scheme to characterize the innovations. First, if the developmental work were "simple," the innovation was characterized as "minor"; conversely, "difficult" development work underlay "major" innovations. Implied value judgments underlying these labels were not completely justified by Hollander's evaluations of the effects of these innovations. Second, if an innovation's primary effect was to increase output for a given level of inputs, it was called "indirect"; if it reduced inputs for a given level of output, it was called "direct." Indirect innovations were primarily methods of increasing the speed of equipment, such as increasing speed of spinning equipment. "Scale" effects were also considered. Plant scale was defined to increase when additional units of equipment were added to the plant, although Hollander recognized that increases in capacity resulting from indirect innovations were also possible. Because capacity gains from indirect innovations were conceptually intertwined with increases in the "scale" of production (as defined by Hollander), his definition of increased scale is objectionable.

Hollander examined the impact of these rayon innovations in an analytical format quite similar to prime and overhead cost analysis. Prime costs were unit labor and unit material costs measured in 1939 prices, thus making them nearly synonymous with labor and material productivities. Unit capital costs were included with other unit costs in unit overhead costs, again measured in 1939 dollars. One finding was that reductions in unit labor cost ranged from 22 to 75 percent across plants, averaging 56 percent, with two-thirds of these reductions attributable to minor, indirect innovations and the remainder to major, direct innovations. Reductions in unit materials costs ranged only from 2 to 30 percent (averaging 16 percent), and were entirely the result of minor, direct innovations. Reductions in unit overhead costs ranged from zero to 75 percent (averaging 42 percent), and resulted from indirect innovations, with more than 70 percent further attributed to minor innovations.

Increases in "scale," which were not uniform across plants,[17] yielded widely varying results: unit labor costs declined between zero and 10 percent; unit materials costs were unchanged in all five plants; and reductions in unit overhead ranged from zero to 50 percent. Increases in plant capacity from all sources were only weakly correlated with total reductions in unit overhead cost on the basis of interplant comparisons. Hollander offered little explanation for this finding except to suggest that it may have been caused by variations in the extent to which fixed assets were depreciated.

Hollander's study was very rich in detail and, for that reason alone, is a valuable addition to the literature. In addition, although we have some reservations about his classification schemes and about the deflation methods used, Hollander was able to demonstrate the effects of several innovations on unit costs more convincingly than has been accomplished in most other studies. He also offered quantitative evidence that minor innovations were cumulatively more important than major innovations in reducing costs, accounting for two-thirds of the reductions in labor and overhead costs and all of the reductions in materials costs.

Studies Oriented Toward the Effects of Technological Change on Labor

Although economists since Adam Smith have been concerned with the impact of technological change on wages and employment, most of the analyses have been conducted at the aggregate level and have culminated in the production function literature summarized in Chapter 2. The literature of general economics has emphasized the aggregate effects because it has been generally oriented toward showing that the direct displacement of labor by a particular machine does not necessarily result in general unemployment and can even lead to an increase in the income of labor as a whole. The general concern with aggregates, however, has meant that most studies of innovation in particular firms and industries have been conducted by investigators with interests narrowly focused on specific economic effects.

One of the most important bodies of such literature falls within the field of labor economics and industrial relations. Because such studies are primarily concerned with labor, they offer only limited contributions to understanding the overall effects of technological change. The explicit concerns of such studies that are also relevant to our own analytical interests include changes in the quantity of employment in the firm or industry undertaking the technological change; the occupational distribution of employment; unit labor requirements as reflected in output per man-hour; and the consequences of "automation," which may be regarded as one form of technological change.

Technological Change and Output per Labor Unit

The Bureau of Labor Statistics (BLS) has long conducted a series of productivity studies, some of which served as the forerunners of the work of Fabricant (1940) and Kendrick (1961).[18] These have been regularly reported in the *Monthly Labor Review* and are a useful starting point for analyzing productivity changes in many industries. The basic approach is to construct indexes of employment and of output for the industry as the basis for an index of output per unit of labor. Where possible, hours of direct and indirect labor are separated. The studies include discussions of technological and other changes affecting the industry, which help in interpreting the findings but are not a formal part of the analysis. Direct measures of the importance of capital, materials, or technology are precluded, of course, by the focus on labor inputs alone. Although physical measures of inputs and outputs are emphasized, prices frequently enter into the construction of output series.

The work by Fuchs (1967) on *Productivity Differences Within the Service Sector* is an extension of the BLS tradition. Fuchs defined output as sales revenue deflated by the relevant components of the consumer or wholesale price indexes; however, for retail trades he measured output as gross margin (sales minus cost of goods sold) times a constant (sales divided by gross margin for 1957-1958). The employment indexes are constructed primarily from employment data, but part-time payrolls are converted to "full-time equivalent employees" by assuming that both kinds of labor received the same hourly wage. The study is of some interest in showing that sales or margin per man changed at different rates in the eighteen service and retail industries, but the interpretation of the findings is subject to all the problems of other one-input studies; in addition, it is based on a measure of output that is only as accurate as the price indexes used to deflate sales and that also ignores the changing role of services provided at retail levels. As is common in industry-level productivity studies, technological change does not enter the analysis specifically.

In the same volume, Wilburn (1967) performs a more detailed analysis, contrasting productivity trends in the barber and beauty shop industries. The primary measurement remains sales per employee, although its significance is somewhat undermined by the author's detailed technological history showing the importance to beauty shops of innovations in capital and materials inputs, as well as changes in the mix of services provided.

The Joint Financial Management Improvement Program (1973) has initiated an attempt to apply productivity analyses of the BLS type at the agency level within the United States government. The main objective of the studies is to improve the productivity of government agencies, and the effort could be productive despite the difficulties involved. These difficulties are chiefly the problems of measuring output in the absence of physical measures or market prices. Even more than the studies of industries in the private sector, therefore, the studies of government productivity are forced to rely on judgments of analysts or managers regarding the meaning of the quantity and quality of output. The Bureau of Customs, for example, processed 2,985,000 parcels mailed from foreign countries to the United States in 1972, compared with 1,549,000 in 1967, an increase of 93 percent. Yet employment in that activity increased only 44 percent, from 511 to 735 man-years (U.S. Joint Financial Management, 1973, p. 11). Does this indicate that output per man-year increased 34 percent, as the data indicate, because of the various managerial improvements recounted in the study? Alternatively, the closeness with which parcels are scrutinized may have decreased. In any event, the output of the agency is obviously difficult to evaluate in economic terms.

Within the private sector, studies at the level of the plant can avoid many of the problems of industry-level studies. In general, the measurement of output will be simpler and technological changes can be dated more specifically, because the diffusion process at the industry level generally spreads the economic impact over a large number of years. The advantages of plant-level data are, however, matched by the outsider's difficulty in gaining access to them, which doubtless explains the paucity of studies.

Youtsler (1972) analyzed annual data for three paper mills in upstate New York for the 1946-1960 period. For each mill, he had data on output (tons), employment of production and nonproduction workers, hourly and weekly earnings of production workers, and direct labor cost per ton. Some information about specific technological changes was also available, but capital and materials inputs were not analyzed. The overall increase in output per man ranged from 38 percent to 200 percent, with direct labor cost per ton decreasing 20 percent in the latter case and increasing 75 percent in the former. But no detailed analysis is provided to indicate the relative contributions of the various factors involved.

Technological Change and Labor Displacement

Much of the literature on the effects of technological change on the quantity and quality of labor is oriented toward, or useful for, industrial relations. Some of the older literature in this area was surveyed by Aronson (1959) who documents numerous references to reduction of direct labor input per unit of output. That the increase in average output per man is not necessarily accompanied by layoffs is indicated by the studies surveyed by Terborgh (1965). Techniques for alleviating the individual hardships of the displaced labor are reviewed in Barkin (1966) and Somers (1963). The three studies summarized below represent three different streams of literature, none of which are primarily concerned with the economic effects of technological change, but all of which present some peripherally relevant data.

Lipstreu and Reed (1964) studied a large bakery before and after its move to a new highly automated plant. Their detailed analysis extended to every aspect of the transition that might be of interest to management except for the economic outcomes, which are treated in only four pages (pp. 143-147). The discussion of such matters as changes in skill mix, morale, supervisory requirements, and interdependence of functions suggests many of the factors that could influence the profitability of particular technological changes. The basic method used in this study was long-term observation within the plant combined with informal conversations with employees and three formal attitude surveys. One can only regret that the superior understanding of the production process provided by this labor-intensive research method was not combined with an equivalent effort to analyze the financial and economic outcomes.

The Stanford Research Institute reported case studies in *Management Decisions to Automate* (1964) that also examined the extent to which managerial expectations of benefits were fulfilled. The eight cases included two banks adopting electronic data processing, two electronic firms adopting numerically controlled machines, and four warehouses adopting automatic order assembly systems. The studies are based on the data obtained from interviews and various documents and records provided by the companies. Examples are provided of savings in the quantity of labor, of changes in the quality of labor and materials, and of the costs of introducing the innovation. Because of the limitations on the amount of information that firms would disclose, however, coherent and complete appraisals of the economic effects of the innovations are not possible.

Kelber and Schlesinger, in *Union Printers and Controlled Automation* (1967), focus on the effects of technological changes (from the introduction of the linotype to the present) on the quantity and quality of labor input and on the reaction of the International Typographical Union to such techniques. Despite their narrow principal focus, the authors provide more comprehensive information about many of the innovations. In particular, the descriptions generally include the manufacturer's estimate of rate of output to be expected from a piece of equipment, the actual operating results, and an explanation of the overall impact on related activities of the newspaper or print shop. Generally, estimates are provided of the initial cost of the equipment and, of course, of the wages of the labor displaced. The strongest feature of the study is the discussion of the attempts by the union to minimize the displacement of its members and to negotiate compensation schedules following the technological changes.

Technological Change and the Quality of Labor

"Quality of labor" has been variously defined in the literature. Most such discussions involve appraisals of the economic effects of technological change, for changes in either the skill mix of the work force or the pleasantness of the work can affect the cost of, and resistance to, technological change as well as average wage rates.

This issue is dealt with in the surveys of literature by Aronson (1959), Terborgh (1965), Stettner (1964), and others mentioned earlier. In general, case studies at the plant level describe some changes in the characteristics of the labor force resulting from technological changes, but it is difficult to compare, or generalize from, the descriptive studies available. A continuing effort to do so at the Institute of Economics of the Czechoslovak Academy of Sciences was described by Auerhan (1966), but unfortunately with insufficient detail to permit any analysis of the method. Solomon Barkin (1966), in discussing the Auerhan paper, suggested that the skill mix could be expected to depend on relative wage rates of different skills rather than being rigidly determined by technology.

Piore, however, suggests that neither the assumption of technically determined manning requirements nor that of responsiveness to wage differentials is correct (1956). In the firms that he studied, the number and skill levels of employees were determined after the introduction of the new equipment by bargaining between unions and management. Furthermore, engineering departments of the firms studied did not take account of wage differentials, having only a somewhat arbitrary single "standard wage rate" to use in estimating savings in the cost of labor. Within the firm, at least, Piore's finding indicates that innovation is not sensitive to relative wages of the various skills or to the exact cost of labor, although increasing output per man-hour remains an objective of technological change. Nevertheless, in studies of the effects of specific innovations, one would still expect to find major interest in reducing total cost and labor cost per unit of output.

Empirical Studies of Specific Innovations

A search of the economics literature yielded only three industries covered by such studies: petroleum refining, ironmaking, and steelmaking. Enos (1958, 1962) investigated four process innovations in petroleum refining; Boylan (1975) and Peirce (1974) recently investigated the effects of ore pelletizing on subsequent operations within the steel industry; and Sethur (1960), Maddala and Knight (1967), and Dilley and McBride (1967) have reported some effects of the basic oxygen steelmaking process on costs.

Petroleum Refining

Enos (1962) studied the effects of thermal cracking, also called the Burton Process (the first commercial plant was built in 1913); polymerization, also called the tube and tank process (the first commercial plant was built in 1922); catalytic cracking, also called the Houdry process (first commercial plant—1937); and catalytic reforming, also called fluid catalytic cracking (first commercial plant—1942). His basic measures of the effects of these innovations encompassed: the average productivities of capital (measured in current dollars), labor (man-hours), materials (volume of crude oil), and energy (BTUs); movements in the prices of these inputs; and changes in the unit cost of capital as well as in total unit cost.

Output was measured in ton-miles of transport hauling by the users of petroleum products to encompass improvements in product quality over time. One major weakness of this measure is that only gasoline output was measured, the volumes and qualities of other distillates being assumed to remain in constant proportions to gasoline.[19] Another weakness of his measure is that some of the increases in ton-miles of transport hauling were patently the result of improvements in vehicle engines and other design features. Enos's capital figures were the amounts of investment required for new plants.

For each of these four innovations, Enos measured the input requirements of the first commercial installation, called the alpha period plant, and of a typical plant installation just before the construction of the first commercial plant embodying a new, presumably superior, process innovation, called the beta period plant. The beta period plant incorporated most of the improvements in the original process including increases in scale. For example, the input requirements in plants employing the thermal cracking process were measured in 1913 (the alpha period) and in 1922 (the beta period); the latter year also representing the alpha period of the successor tube and tank process.

If the unit input requirements of successive innovations are compared in their alpha periods or in their beta periods, in almost every case all four inputs were saved.[20] Also, these four innovations had a large cumulative impact. For example, a beta period comparison of the first innovation (thermal cracking) to the fourth (catalytic reforming)—that is, comparing a typical 1922 plant to a typical 1955 plant—indicates the unit labor requirement fell by 95 percent, the unit capital requirement by 50 percent in current dollars, the unit energy requirement by 85 percent, and the unit materials requirement by 60 percent.

The least favorable method of contrasting a new technology with its predecessor is to compare the unit input requirements of the alpha period plant of the new technology with those of a beta period plant of the predecessor technology. This would be one important basis for deciding whether to invest in the alpha period plant, provided that management did not expect subsequent major improvements in the new technology. The results of this comparison are less in accord with expectations, inasmuch as unit input requirements rose in half the cases. This finding reflects Enos's appraisal that continuing improvements in an innovation were at least as important as the initial innovation itself (that is, as reflected in the first commercial plant) in saving inputs, and may indicate that the first commercial plant should be considered part of the process of developing an innovation into a commercial success. In fact, Enos believed that firms expected large improvements in plant design and operating characteristics to occur during beta periods.[21]

Because the overall trend in input saving was greatest for labor and least for capital throughout the 1913-1955 period, Enos attempted to correlate these disparate savings with earlier changes in relative factor prices. The correlation, however, was weak—with the possible exception of labor versus the other factors taken as a group. This indicates, according to Enos, that changes in factor proportions were induced by the innovations, rather than planned in response to changes in relative factor prices.

In an earlier account, Enos (1958) found the annual return on undepreciated investment for alpha stage plants to range from 26 percent before tax for polymerization to 49 percent before tax for catalytic reforming. Each innovation was shown to increase revenue per barrel of processed crude oil more than cost per barrel. Reasons for increased revenue per barrel included a shift to "higher" level distillates (such as gasoline), improvements in the qualities of the higher level distillates (such as an increased octane rating of gasoline) and greater product mix flexibility. However, because these relatively high profitability figures belie the modest productivity improvements achieved when a new process was commercially introduced in an alpha stage plant, as discussed above, we are forced to conclude that either the data were imprecise or the revenue-improving characteristics of each innovation were not totally captured by Enos's measure of output.

During the beta phase of development, he found that the unit capacity cost (in constant dollars) fell by at least 64 percent (for catalytic reforming between 1949 and 1952) and as much as 78 percent (for thermal cracking between 1913 and 1946), suggesting that profitability was at least as high here as it was for alpha phase innovators.

Pelletizing and the Production of Iron

In the blast furnace sector of the steel industry, Boylan (1975) estimated the impact of improved iron materials on the productivities of capital, labor, energy, and materials. The major improvements in iron materials were achieved by agglomerating fine particles of ore into sinter and by beneficiating low grade ores to produce pellets.

Pelletizing was developed to the point of commercial success in response to the dwindling domestic supplies of ores that were acceptable in their natural state as a blast furnace material. Pelletizing technology is basically a two-step process in which low-grade taconite ores (such as those with an iron content of 25 percent) are first concentrated and then agglomerated to produce a blast furnace feed with an iron content over 60 percent and also with superior physical properties.

Boylan estimated that pellets doubled the productivities of capital and labor within the blast furnace sector by doubling the throughput of a blast furnace of given size. In addition, less iron materials were wasted and energy productivity was improved by at least 10 percent. Pelletizing also substantially increased the possible size of new blast furnaces, paving the way for further reductions in total unit cost, although the U.S. industry was slow to take advantage of this effect because the existing stock of furnaces was more than adequate to supply the iron needed by the steel industry.

Sintering was first adopted commercially in 1913 to enable blast furnace operators to recycle fine particles of ore and coke, called flue dust, blown out of the blast furnace. Flue dust was produced as a by-product after it had become economically attractive to capture and clean blast furnace gas (for example, to heat the blast furnace stoves). Sintering was used primarily as a method of recycling flue dust until the late 1940s. Subsequently, an increasing number of plants adapted this process to utilize increasing proportions of natural ore fines that were high in iron content (55 to 65 percent) but physically unacceptable in their natural state. Sinter achieved its major growth and commercial success from 1955 to 1965. Sinter quality gradually improved such that high grade sinter had yielded operating results equivalent to those derived from pellets by 1970.

Despite extended theoretical discussions, such as Fellner (1962), attempts to show that technological innovations seek to change factor proportions in response to changes in relative factor prices have met with little success generally. They also failed in this particular case, partly because the effects of other innovations in blast furnace practice were mixed together with the effects of pellets and sinter.

Peirce (1974) investigated the broader effects of pelletizing. He estimated that the capital invested in pelletizing facilities in the United States was $1.2 billion by 1969. But approximately the same amount of capital had been saved through the doubling of blast furnace productivity. Moreover, he surmised that, if pelletizing had not succeeded, the United States would have been forced to rely much more substantially on foreign ore supplies, thereby necessitating greater investment not only in blast furnaces and related equipment, but also in foreign mines, ocean transport, and port handling facilities. Hence, pelletizing in reality saved capital.

Peirce also identified a series of successive effects (or "ripples") from pelletizing. The first ripple reinforced the traditional ways of doing things in the steel industry by safeguarding the economic viability of the major steel plants along the Great Lakes and of the lake fleet along with related transportation facilities. The improved productivity of blast furnaces also greatly reduced pressures for developing technologies to bypass the blast furnace process via some form of direct reduction. A second ripple altered the transportation network for iron ore. Because pellets are very uniform in size and do not freeze together in the winter, they made it possible to design larger lake vessels with self-unloading devices, with resulting shorter unloading times and with the ability to operate through a longer season. Because pellets do not freeze together, interest in unit trains as a method of year-round (continuous) transportation has also been revived.

Peirce envisions a third set of possible ripples in the future, including a gradual consolidation in the number of totally integrated steelmaking plants induced by the increased minimum efficient scale of blast furnaces; additional investment in port facilities and lake ships to take advantage of the economies of handling pellets by conveyor belts; and the construction of pipe lines to transport a slurry of concentrated iron ore to blast furnace plants—the inherently cheapest method of transporting such material.

Steelmaking and the Basic Oxygen Process

Sethur (1960) estimated the savings in total unit cost attributable to the adoption of the basic oxygen process in Austria by comparing actual costs during the 1950s to what such costs would have been if the open hearth method had continued to be employed in a plant of comparable size. He estimated a 10 percent reduction for basic oxygen steelmaking in the cost of assembling the quantities of hot iron, steel scrap, and oxygen needed per ton of steel. Austria had ample blast furnace capacity but a shortage of steel scrap; also its basic oxygen furnace used a higher ratio of hot iron to scrap. He also estimated a 47 percent reduction in the total cost per ton of converting the above inputs into steel, mostly because of lower material and energy costs (which dominate conversion cost). But labor and capital amortization costs per unit were also reduced by 35 to 40 percent.

Maddala and Knight (1967) used *hypothetical* cost data for steel plants in Latin America, Mexico, and Europe in the early 1960s to determine the conditions under which basic oxygen steelmaking would be more efficient than electric or open hearth furnaces. These conditions were scrap scarcity, excess pig iron capacity, capital limitations, and a narrow range of desired steel products (such as carbon strip). The total unit cost of the basic oxygen method was estimated to be less than the open hearth method by 12 percent in Latin America, by 17 percent in Mexico, and by 3 to 20 percent in Europe, depending on relative input prices. More specifically, unit labor cost was lower by 33 percent in Latin America, by 40 percent in Mexico, and 44 percent in Europe. Unit capital cost was lower by 33 percent in Latin America and Mexico and by 50 percent in Europe.

For every unit cost comparison made above, an electric furnace plant fared better than an open hearth plant, but the basic oxygen process was still judged to be cheaper, except in Europe, when the price of iron ore was assumed to be relatively high, scrap prices relatively low, and all other input prices relatively "medium." Under these input price assumptions, Maddala and Knight reported that electric furnaces could be as much as 10 percent cheaper than basic oxygen converters in producing carbon steel. Finally, Maddala and Knight estimated economies of scale to be important but equal in all three processes, thus eliminating this factor in choosing among these processes.

Dilley and McBride (1967) examined diffusion of basic oxygen steelmaking in the United States, but also provided some interesting cost estimates. Commenting on a Kaiser Engineers' study made in 1955, that the total unit cost of basic oxygen steel was \$5 to \$6 per ton cheaper than open hearth steel for a plant capacity ranging from 500,000 to 1,000,000 tons, they estimated that if the capital costs of the open hearth facility were treated as "sunk" (because the open hearth capacity was already installed), this advantage would be reduced to \$0.75 per ton or less, making it difficult to justify scrapping proven open hearth capacity for basic oxygen capacity that was far more restricted in its currently feasible scale and in its ability to produce the variety of steels demanded in the United States. They also claimed that the estimated operating cost advantage of the basic oxygen process for United States Steel (their employer) was well below the estimate made by Kaiser Engineers. Finally, they noted that the Kaiser estimate did not take account of additional investment in blast furnaces that would have been necessitated by the adoption of the new basic oxygen process. (By 1960, the increasing importance of pellets would have greatly reduced the need for such additional investment.) Only after U.S. steel firms improved the basic oxygen process by increasing the scale of converters and by expanding the quality and range of steels that could be produced, did this process prove to have a clear-cut cost advantage over the open hearth process, according to Dilley and McBride. But they offered no comparative cost figures concerning the size of this advantage.

Automation Effects

Oddly enough, one of the most comprehensive studies of the effects of actual technological changes was published quite early. James Bright (1958) analyzed thirteen cases of the introduction of "more automatic" production systems covering a wide range of industries. Because of his primary concern with the implications of automation for managements, measures of specific economic effects received little emphasis. But Bright points out that, aside from their reluctance to divulge information about attendant cost changes, managements were in fact often unable even to compute them because of the way that basic records were kept and because of the large number of changes occurring simultaneously with such innovations.

Direct observation of the new equipment combined with extensive interviews enabled Bright to piece together information about the objectives of managements in making the changes and about the outcomes. Bright observed that "no firm was completely successful in anticipating all the major—let alone the minor—advantages and disadvantages that followed automation," and that "managements tend to evaluate proposals within a rather limited framework" (p. 82). Even without statistical data, he was able to determine that reductions in direct labor input were generally one of the primary objectives and were generally achieved. On the less favorable side, however, unanticipated cost increases were the rule rather than the exception (p. 115).

Bright's analysis included both advantages and disadvantages. The former included increased capacity; reduced direct and indirect labor; shorter lead time; higher quality; scrap reduction; improvements in safety, morale, and working conditions; and increased product flexibility. Disadvantages included troublesome and expensive start-up period; reduced flexibility in inputs, outputs, and scheduling; and the need for faster administrative reaction time. Although these lists are suggestive rather than exhaustive, they illustrate the necessity of taking a broader view of the effects of innovation than has been common in economic analyses. They also emphasize both the difficulty and the importance of capturing the numerous qualitative changes associated with changes in production techniques.

The Diffusion of Innovations

The economic effects of new technologies obviously depend on their rates of entry into commercial use. Studies of such diffusion rates and of the factors affecting differences among them might be expected to have a bearing on policy issues relating to the desirability of increasing or decreasing such rates and to the means of effectuating such intentions.

Models and Measures Used

Most studies of the diffusion of innovations by economists[22] and by other social scientists as well[23] have assumed an S-curve model and have then concentrated on estimating its parameters (and the determinants of these parameters) for the particular innovation covered. Results have ranged from relatively strong to relatively weak confirmation of such patterns,[24] using measures either of the proportion of potential adopters using the innovation or of the proportion of total relevant output accounted for by the innovation.[25] As Rosegger noted in Chapter 4, the latter tends to be superior for most economic analyses because merely counting the number of users does not indicate the extent of their changeover to the innovation.

But much more serious questions can be raised about the usefulness of S-curves in analyzing diffusion patterns. Nabseth argues that an important objective of uncovering diffusion patterns is to be able to forecast future diffusion rates from results during the early years of adoption. In this connection, however, he finds that empirical findings yield so many S-curve patterns that a simple linear extrapolation is likely to yield comparable, if not superior, predictions. An even more troublesome question concerns the economic interpretation to be made of S-curve results. Lacci, Davies, and Smith (1974) have suggested that estimates of the rate of returns from adopting an innovation tend to rise over time among potential adopters because of increasing information about the technical and financial aspects of the innovation. Others who have employed information flows as an explanatory variable include Griliches (1957) and Mansfield (1961) among early users and Bernhardt (1970) among more recent studies. But it will be recalled that Rosegger argued knowledgeably in Chapter 4 that external information plays a relatively minor role in decisions by industrial firms to adopt major innovations.

Thus, the alternative conception must be considered that *ex post* delineations of S-curve patterns imply a spurious homogeneity to the actual experiences encompassed by such aggregate results. Specifically, Gold, Peirce, and Rosegger (1970) suggest that detailed examinations of technological diffusion in industry reveal successive changes in the nature of the innovation (including its reliability, capabilities, efficiency, and scale), in factor and product market pressures affecting the attractiveness of the innovation, and in the availability of financial resources and technical expertise in firms contemplating its adoption. Nabseth acknowledged awareness of this broader viewpoint (1973, p. 257, note 2) and later restated it (1974, p. 303); Ray (1969) expressed a similar view by noting that the innovation adopted by pioneers often differs materially from those adopted by the laggards. Accordingly, the S-curve would seem to have little analytical usefulness for there would seem to be little basis for differentiating between actual adopters and potential adopters except by ignoring factors that nonadopters at any given period might consider a sufficient basis for rejecting favorable action. Accordingly, attention may turn more usefully to studies of the factors affecting differences in diffusion rates.

Factors Affecting Diffusion Rates

Differences in diffusion rates tend to be attributed to human factors and to economic factors. The former, including managerial receptivity to change and to risk-taking as well as managerial aggressiveness vis-à-vis competitors, are generally not amenable to objective measurement, although they are undoubtedly important and subject to a wide range of differences among companies, as has been noted by Gold (1964). Accordingly, most studies, especially those by economists, tend to concentrate on economic factors and to cite human factors only in seeking to account for residual differences (see, for example, Meyer and Herregat, 1974).

The magnitude of expected profit rates is generally regarded as the most important cause of differences in diffusion rates but direct estimates have proven to be difficult to obtain *ex post* as well as *ex ante*, leaving this an article of faith rather than an empirically demonstrated fact.[26] Other factors that have been proposed as explanatory variables may be classified into three groups: technological characteristics (products, processes, and innovations), characteristics of the firm, and characteristics of the market.

Technological Factors. The number of firms in an industry which are potential adopters of an innovation, and the proportion of their output to which it might be applied, depends on the functional specificity of the innovation at successive stages of development as well as on the range of relevant processes and products in individual plants. Hence, adoption rates cannot properly be compared with the total number of firms in, or the total output of, their common "industry" classification. Rather, the progressively changing characteristics of the innovation in its various forms must be accompanied by changing measures of the array of economically feasible applications. Moreover, identification of the constraints on more rapid diffusion would help to differentiate the role of technological limitations as compared with economic considerations, with obvious relevance to policies concerned with accelerating diffusion.

Such technological restrictions on diffusion are illustrated for basic oxygen steelmaking by Maddala and Knight (1967), Dilley and McBride (1967), and Meyer and Herregat (1974). Similar technological restrictions are noted by Schenk (1974) in respect to the continuous casting of steel, by Smith (1974) in respect to water jet shuttleless looms, Peirce (1974) in respect to the cleaning and continuous mining of coal, and Ray (1974) in respect to the float glass process.

The reverse pattern of a change in market demand tending to encourage adoption of a new technology is reported by Peirce (1974), in respect to the increased adoption of mechanical techniques in coal mining after World War II because of the major shifts in market demand from lump coal to crushed coal. And still another form of this market pressure is represented by governmental safety and pollution control and other pressures which force shifts in the technological characteristics sought by buyers. For example, Gold, Peirce, and Rosegger (1970) mention the role of government restrictions on tin consumption in accelerating the diffusion of electrolytic tin plating. Lacci, Davies, and Smith (1974) report that adoption of the tunnel kiln in England was encouraged by the hazards of the earlier process; Ray (1974) found that the use of gibberallic acid in brewing was prohibited in Germany; and Peirce (1974) expects a sharp increase in the diffusion of mechanical coal cleaning if its removal of sulfur should be significantly improved.

Diffusion rates also tend to be restrained by the extent to which adoption of an innovation involves disrupting the integrated network of surrounding activities or the balance among their respective productive capacities. Such additional effects of the innovation's technological characteristics are also reported by the steel studies cited above as well as by Sutherland (1959) in connection with cotton spinning innovations and by Ray (1974) in respect to the diffusion of gibberallic acid in brewing. Conversely, innovations that help to redress imbalances among production stages are likely to engender faster diffusion, as noted by Peirce (1974) in respect to the mutual reinforcement of pelletizing iron ore and adopting basic oxygen steelmaking, the former increasing the pig iron capacity of blast furnaces at the same time as the latter increased the demand for it.

Still another technological factor affecting diffusion rates relates to the risk of unreliability in the functioning of the innovation. Management perceptions of such risk combine both psychological reactions to risk-taking and technical evaluations of operational uncertainties. Concentrating for present purposes on the latter, Mansfield (1961, 1963) found that such technical evaluations of risk affected adoption decisions significantly in by-product coke ovens and in three other industries, as did Lacci, Davies, and Smith (1974) in tunnel kilns for brickmaking and Schenk (1974) in the continuous casting of steel.[27] However, eliciting such judgments via questionnaires after the innovation has "succeeded" yields only highly subjective results at best. Finally, the diffusion of innovations has also been slowed by improvements in the old technology that reduce the net advantage of the new, as has been noted by Peirce (1974) and Lacci, Davies, and Smith (1974).

Firm Characteristics. The characteristics of firms that seem to influence the diffusion rates of innovations include the scale of their operations, the age distribution of existing capital facilities, and the availability of additional investment funds. The large scale of dominant plant operations in the United States relative to the small scale of the basic oxygen furnaces and continuous casting machines available during the early applications of these innovations has been shown by Dilley and McBride (1967) and by Huettner (1974) to help explain the slow diffusion for which the steel industry was attacked by Adams and Dirlam (1964) and by Ault (1973). The reverse disparity between large capacity machines and relatively small firms was noted by Peirce (1974) in explaining the slow diffusion of mechanical techniques in coal mining and Ray (1974) reported that the enormous scale of float glass processes forced delay in its adoption in continental Europe until consolidation resulted in control of the market by only three companies. Between these extremes, the diffision of numerically controlled machine tools was found to be most rapid, both by Gebhardt and Hatzold (1974) and by Romeo (1975), in firms specializing in small to medium-sized batch operations.

Firms with relatively recently acquired facilities representing the prior technology tended to adopt new technologies more slowly than firms with older, and presumably more fully depreciated facilities according to the findings of Ray (1974) in respect to float glass, Mansfield (1963) in respect to diesel locomotives, Lacci, Davies, and Smith (1974) in regard to tunnel kilns, and Dilley and McBride (1967) in respect to basic oxygen steelmaking. Empirical findings are in less agreement, however, about the effect of expanding output levels on diffusion rates. The above-cited studies by Ray, Lacci et al., Meyer and Herregat, and Dilley and McBride find diffusion rates tending to increase with market expansion, although this connection was not strong, whereas Gold, Peirce, and Rosegger (1970) found no consistent relationship between them during the first fifteen years of commercial application of fourteen major innovations in the iron ore, coal and steel industries. Nabseth and Ray (1974) have attempted to generalize these effects by suggesting that market growth underlies the accelerator theory of investment and the age distribution of capital stock reflects the vintage approach to investment theory, such as Salter's model (1960)—both these variables being considered as imperfect proxies for profitability in Nabseth's view. It might be added, in this connection, that there has been widespread recognition that diffusion rates would also tend to be inhibited by inadequate financial resources on the part of prospective adopters.

Reductions in Cost and Improved Profitability. Although it has proved very difficult to estimate the distinctive effects of innovations on profitability, both *ex ante* and *ex post*, as noted earlier, a number of studies have linked diffusion rates to profitability, including Mansfield (1963a, 1963b), Romeo (1975), Ray (1969), and Griliches (1957). Others have emphasized the importance of factor prices and their presumed affect on profitability, including the steel studies by Meyer and Herregat (1974) and by Dilley and McBride (1967). Oddly enough, very few studies have been found relating diffusion rates to the relative strength of trade unions. A few of the studies in Nabseth and Ray (1974) reported that labor-saving innovations were adopted more rapidly in the United States than in Europe and attributed this to relatively higher wage rates in the United States, although without confirmatory correlation analyses. Such views are also supported by Peirce's findings (1974) that mechanical techniques in coal mining were adopted more rapidly in highly unionized mines.

Despite the paucity of unambiguous supporting evidence, it is generally believed that expected profitability is the most important determinant of diffusion rates[28]—as well as of all other voluntary managerial decisions in the private sector. But this would seem merely to emphasize the need to dispel our ignorance about how managements arrive at the profit expectations determining their decisions and to what extent, and by what means, they evaluate the outcomes of such decisions.

Notes

1. The indirect definition partly overlaps commonly used measures of technological change.

2. Mansfield defined technology as "society's pool of knowledge regarding the industrial arts. It consists of knowledge used by industry regarding the principles of physical and social phenomena . . . knowledge regarding the application of these principles to production . . . and knowledge regarding the day-to-day operations of production" (1968, pp. 10-11).

3. A more commonly accepted distinction between these two concepts is presented in Chapter 4.

4. See the next section. For example, if a new machine is developed by a machine manufacturer, it would be studied as a process innovation in the industries employing the new machine, rather than studied as a new product of the machine tool industry.

5. If the production process is characterized by constant returns to scale, both before and after technical change has occurred, then it is not necessary to specify the level of output.

6. In this simple case, the capital to labor ratio at point B compared to point A would determine whether the innovation saves relatively more capital, labor, or is neutral.

7. An important qualification is that Gold does not accept the isoquant concept, finding it much too simplistic to be of much help in production theory.

8. Of course, obsolescence is normally hastened by the development of the innovation. Salter's model (1960) states that old facilities will be replaced when the total unit cost of the innovation is below average variable cost in old facilities, implying a monopolizing concern with cost effects, which is in striking contrast to the analysis in Gold, Peirce, and Rosegger (1970).

9. These three effects can be illustrated via the unit isoquant approach presented earlier as (a) the fall in the unit isoquant measured along the same capital to labor ratio that prevailed prior to the innovation; (b) the capital-labor substitution induced by the change in the shape of the unit isoquant but with unchanged factor prices; and (c) the movement along the new unit isoquant induced by a change in factor prices.

10. Or else a smaller increase in employment plus an increase in the wage rate would occur, up to a maximum of no increase in employment and a 1 percent increase in wages for each 5 percent increase in government expenditures.

11. Petersen also measured capacity by employing the same index number measure as he used for output. Thus, he was able to use a better measure of capital productivity (capacity divided by net fixed investment) than Johnson used (output divided by net fixed investment). Capacity utilization fluctuated from 80 to 100 percent of its 1947 level during the 1947-1963 period, but reached the 1947 level in 1963. Thus, this difference in the measures of capital productivity used would not account for any differences in net change for the period covered by Petersen's study.

12. In fact, materials prices rose 50 percent of the time while, on the other hand, wage rates rose 85 percent of the time—closely paralleling the frequency of reductions in unit labor requirements (75 percent).

13. Measured by pounds of cloth produced per man-hour at the plant level, and by value added in constant 1954 dollars per employee at the industry level.

14. Capacity in 1954 dollars divided by net fixed investment in 1954 dollars and output/horsepower of power equipment where output was measured as value added in 1954 dollars or pounds of cloth.

15. Output/kilowatt hours of electricity consumed.

16. The four subperiods were 1904-1921, 1921-1929, 1929-1947, and 1947-1969. Capital productivity was measured by capacity ÷ net fixed investment, labor productivity by output per man-hour, and the capital-labor ratio by actively utilized net fixed investment ÷ total man-hours, where actively utilized fixed investment is determined by multiplying net fixed investment by the fraction of total capacity actually utilized.

17. Hollander did not break down expansions in plant capacity into expansions attributable to indirect innovations and expansions resulting from "scale."

18. Ewan Clague (1966) traces the inception of productivity measurement to Carroll D. Wright's work in the 1890s.

19. Enos noted that the volume of these other distillates exceeded the volume of gasoline in every year. He was not able to measure their quality improvements, but did mention their aggregate value ranged from one-half to two-thirds of the value of gasoline from 1913 to 1955.

20. The two exceptions occurred when the catalytic cracking (Houdry) process was compared to the catalytic reforming (fluid catalytic cracking) process: in the alpha period comparison, the energy requirement was increased; and in the beta comparison, the capital requirement was increased.

21. But this raises serious questions concerning what is defined as "best practice," at what stage to adopt major innovations, and what the critical decision criteria should be.

22. Early contributions include those of Griliches (1957), Sutherland (1959), and Mansfield (1961).

23. For example, see Ryan and Gross (1943).

24. For example, Griliches (1962) and Mansfield (1961) had results favoring the *S*-curve, while Sutherland's results (1959) were less favorable.

25. Because he studied major innovations that involved the development of new industries, Lynn (1966) shifted to measuring diffusion in terms of the percentage of GNP accounted for by each innovation.

26. See Nabseth and Ray (1974, pp. 301-306).

27. Similar retardation of diffusion because of the small size of firms is reported by Lacci, Davies, and Smith (1974) and Sutherland (1959) as well as by Smith (1974) in respect to shuttleless looms and by Metcalfe (1970) in respect to power looms.

28. Ray's summary of the factor's affecting diffusion rates ranked them in the following order of importance: profitability; management attitudes; and access to needed capital—but hard evidence has been very largely missing in respect to all three.

References

Adams, W., and J. Dirlam, "Steel Imports and Vertical Oligopoly Power," *American Economic Review* (September 1964).

_______, "Big Steel, Invention and Innovation," *Quarterly Journal of Economics* (May 1966).

Aronson, R., "Automation—Challenge to Collective Bargaining," in *New Dimensions in Collective Bargaining*, ed. H. Davey, H. Kaltenborn, and S. Ruttenberg (New York: Harper, 1959).

Auerhan, J., "Technological Change in the Socialist Countries," in *Employment Problems of Automation and Advanced Technology: An International Perspective* (London: Macmillan, 1966).

Ault, D., "The Continued Deterioration of the Competitive Ability of the U.S. Steel Industry: The Development of Continuous Casting," *Western Economic Journal* (March 1973).

Barkin, S., "Programming of Technical Change and Manpower Adjustments," *Employment Problems of Automation* (1966).

Bauer, I.L.L., "The Effect of Technology on the Farm Labor Market," *American Journal of Agricultural Economics* (August 1964).

Bernhardt, I., "Diffusion of Catalytic Techniques Through a Population of Medium Size Firms," *The Journal of Industrial Economics* (November 1970).

Blaug, M., "A Survey of the Theory of Process Innovations," *Economica* (February 1963).

Boylan, M., *Economic Effects of Scale Increases in the Steel Industry: The Case of U.S. Blast Furnaces* (New York: Praeger, 1975).

Bright, J., *Automation and Management* (Boston: Harvard University, Graduate School of Business Administration, 1958).

Brown, M., *On the Theory and Measurement of Technological Change* (Cambridge: Cambridge University Press, 1968).

Butcher, W., "Productivity, Technology, and Employment in U.S. Agriculture," in *Studies Prepared for the National Commission on Technology, Automation and Economic Progress*, Appendix, vol. 2 (Washington, D.C.: U.S. Government Printing Office, 1966).

Chow, G., "Technological Change and the Demand for Computers," *American Economic Review* (December 1967).

Clague, E., "Measurement of Technological Change," in *Employment Problems of Automation* (1966).

Coleman, J., E. Katz and H. Menzel, "The Diffusion of an Innovation Among Physicians," *Sociometry* (December 1957).

Davey, H., H. Kaltenborn, and S. Ruttenberg (eds.), *New Dimensions in Collective Bargaining* (New York: Harper, 1959).

Dilley, D., and D. McBride, "Oxygen Steelmaking–Fact vs. Folklore," *Iron and Steel Engineer* (October 1967).

Enos, J., "A Measure of the Rate of Technological Progress in the Petroleum Refining Industry," *Journal of Industrial Economics* (June 1958).

_______, "Invention and Innovation in the Petroleum Refining Industry," in *The Rate and Direction of Inventive Activity: Economic and Social Factors*, ed. R. Nelson (Princeton, N.J.: Princeton University Press, 1962).

_______, *Petroleum Progress and Profits: A History of Process Innovation* (Cambridge, Mass.: MIT Press, 1962).

Fabricant, S., *The Output of Manufacturing Industries, 1899-1937* (New York: National Bureau of Economic Research, 1940).

Fellner, W., "Does the Market Direct the Relative Factor Saving Effects of Technological Progress?" in *The Rate and Direction of Inventive Activity* (1962).

Fuchs, V., and J. Wilburn, *Productivity Differences Within the Service Sector*, National Bureau of Economic Research, Occasional Paper 102 (New York: Columbia University Press, 1967).

Gebhardt, A., and O. Hatzold, "Numerically Controlled Machine Tools," in *The Diffusion of New Industrial Processes: An International Study* (London: Cambridge University Press, 1974).

Gold, B., *Foundations of Productivity Analysis* (Pittsburgh: University of Pittsburgh Press, 1955).

_______, "Economic Effects of Technological Innovations," *Management Science* (September 1964).

_______, "Technology, Productivity and Economic Analysis," *OMEGA* (February 1973a).

_______, "The Impact of Technological Innovation–Concepts and Measurement," *OMEGA* (April 1973b).

_______, W. Peirce, and G. Rosegger, "Diffusion of Major Technological Innovations in U.S. Iron and Steel Manufacturing," *Journal of Industrial Economics* (July 1970).

_______, "Productivity Analysis: Some New Analytical and Empirical Perspectives," *Business Economics* (May 1974).

Griliches, Z., "Hybrid Corn: An Exploration in the Economics of Technical Change," *Econometrica* (October 1957).

_______, "Research Costs and Social Returns: Hybrid Corn and Related Innovations," *Journal of Political Economy* (October 1958).

Hakanson, S., "Special Presses in Paper-Making," in *The Diffusion of New Industrial Processes* (1974).

Hicks, J., "Distribution and Economic Progress: A Revised Version," *Review of Economic Studies* (October 1936).

Hollander, S., *The Sources of Increased Efficiency: A Study of DuPont Rayon Plants* (Cambridge, Mass.: MIT Press, 1965).

Huettner, D., "The Development of Continuous Casting in the U.S. Steel Industry: Comment," *Economic Inquiry* (June 1974).

Jerome, H., *Mechanization in Industry* (New York: National Bureau of Economic Research, 1934).

Johnson, R., *Railroad Technology and Manpower in the 1970's*, U.S. Dept. of Labor Bulletin No. 1717 (Washington, D.C.: U.S. Government Printing Office, 1972).

Kelber, H., and C. Schlesinger, *Union Printers and Controlled Automation* (New York: The Free Press, 1967).

Kendrick, J., *Productivity Trends in the U.S.* (Princeton, N.J.: Princeton University Press, 1961).

Kennedy, C., and A. Thirlwall, "Surveys in Applied Economics: Technical Progress," *Economic Journal* (March 1972).

Lacci, L., S. Davies, and R. Smith, "Tunnel Kilns in Brick-Making" in *The Diffusion of New Industrial Processes* (1974).

Lipstreu, O., and K. Reed, *Transition to Automation: A Study of People, Production and Change* (Boulder: University of Colorado Studies, 1964).

Lynn, F., "An Investigation of the Rate of Development and Diffusion of Technology in Our Modern Society," in *Studies Prepared for the National Commission on Technology* (1966).

McAdams, A., "Big Steel, Invention and Innovation Reconsidered," *Quarterly Journal of Economics* (August 1967).

Maddala, G., "Productivity and Technical Change in the Bituminous Coal Industry, 1919-1954," *Journal of Political Economy* (August 1965).

_______, and P. Knight, "International Diffusion of Technical Change: A Case Study of the Oxygen Steel Making Process," *Economic Journal* (September 1967).

Mansfield, E., "Technical Change and the Rate of Imitation," *Econometrica* (October 1961).

_______, "The Speed of Response of Firms to New Techniques," *Quarterly Journal of Economics* (May 1963a).

_______, "Intrafirm Rates of Diffusion of an Innovation," *Review of Economics and Statistics* (November 1963b).

_______, "Technological Change: Measurement, Determinants, and Diffusion" in *Studies Prepared for the National Commission on Technology* (1966).

_______, *The Economics of Technological Change* (New York: Norton, 1968).

Metcalfe, J., "Diffusion of Innovation in Lancashire Textile Industry," *Manchester School* (June 1970).

Meyer, J., and G. Herregat, "The Basic Oxygen Steel Process," in *The Diffusion of New Industrial Processes* (1974).

Nabseth, L., and G. Ray (eds.), *The Diffusion of New Industrial Processes: An International Study* (London: Cambridge University Press, 1974).

Nelson, R., (ed.), *The Rate and Direction of Inventive Activity: Economic and Social Factors* (Princeton: Princeton University Press, 1962).

_______, M.J. Peck, and E. Kalachek, *Technology, Economic Growth and Public Policy* (London: George Allen and Unwin, 1967).

Peirce, W., "The Ripple Effects of Technological Innovation: The Case of Iron Ore Pelletizing," *OMEGA* (February 1974).

_______, "Factors Affecting Responsiveness to Technological Innovations in Coal Mining," Working Paper No. 54, Research Program in Industrial Economics, Case Western Reserve University (1974).

Petersen, D., "Economic Effects of Technological Innovations in Railroads, 1947-1963," Working Paper No. 15, Research Program in Industrial Economics, Case Western Reserve University (1968).

Phillips, A., "Concentration, Scale and Technological Change in Selected Manufacturing Industries 1899-1939," *Journal of Industrial Economics* (June 1956).

Piore, M., "The Impact of the Labor Market on the Design and Selection of Productive Techniques Within the Manufacturing Plant," *Quarterly Journal of Economics* (June 1956).

Ray, G., "The Diffusion of New Technology: A Study of Ten Processes in Nine Industries," *National Institute Economic Review* (May 1969).

_______, "Float Glass," in *The Diffusion of New Industrial Processes* (1974).

_______, "The Application of Gibberellic Acid in Malting," in *The Diffusion of New Industrial Processes* (1974).

Rogers, E., *The Diffusion of Innovations* (New York: Free Press of Glencoe, 1962).

Romeo, A., "Interindustry and Interfirm Differences in the Rate of Diffusion and Innovation," *Review of Economics and Statistics* (August 1975).

Rosegger, G., "Steel Imports and Vertical Oligopoly Power: Comment," *American Economic Review* (September 1967).

Rosenberg, N., "Factors Affecting the Diffusion of Technology," *Explorations in Economic History* (Fall 1972).

Ruttan, V., "Usher and Schumpeter on Invention, Innovation and Technological Change," *Quarterly Journal of Economics* (November 1959).

Ryan, B., and N. Gross, "The Diffusion of Hybrid Seed Corn in Two Iowa Communities," *Rural Sociology* (March 1943).

Salter, W., *Productivity and Technical Change* (Cambridge: Cambridge University Press, 1960).

Schenk, W., "Continuous Casting of Steel," in *The Diffusion of New Industrial Processes* (1974).

Schumpeter, J., *Business Cycles* (New York: McGraw-Hill, 1939).

Sethur, F., "Austrian Steel: Development and Prospects," *Journal of Industrial Economics* (June 1960).

Smith, R., "Shuttleless Looms," in *The Diffusion of New Industrial Processes* (1974).

Solow, R., "Technical Change and the Aggregate Production Function," *Review of Economics and Statistics* (August 1957).

Somers, G., E. Cushman and N. Weinberg, *Adjusting to Technological Change* (New York: Harper & Row, 1963).

Stanford Research Institute, *Management Decisions to Automate* (Washington, D.C.: U.S. Manpower Administration, 1964).

Steiber, J., *Employment Problems of Automation and Advanced Technology: An International Perspective* (London: Macmillan, 1966).

Stettner, L., "Survey of Literature on Social and Economic Effects of Technological Change," in *Transition to Automation* (1964).

Studies Prepared for the National Commission on Technology, Automation and Economic Progress, Appendix, Volume 2 (Washington, D.C.: U.S. Government Printing Office, 1966).

Sutherland, A., "Diffusion of an Innovation in Cotton Spinning," *The Journal of Industrial Economics* (March 1959).

Terborgh, G., *The Automation Hysteria* (Washington, D.C.: Machinery and Allied Products Institute, 1965).

Tilton, J., *The International Diffusion of Technology: The Case of Semiconductors* (Washington, D.C.: The Brookings Institute, 1971).

U.S. Joint Financial Management Improvement Program, *Measuring and Enhancing Productivity in the Federal Government* (Washington, D.C.: Office of Management and Budget, June 1973).

Weinberg, E., *Impact of Technological Change and Automation in the Pulp and Paper Industry*, U.S. Dept. of Labor Bulletin 1347 (Washington, D.C.: U.S. Government Printing Office, 1962).

Williams, B.R., (ed.), *Science and Technology in Economic Growth* (London: Halstead Press, 1973).

Youtsler, J., "Automation–Its Effects on the Worker in the Paper Industry," *Paper Trade Journal* (August 6, 1972).

Zeisel, R., *Technology and Manpower in the Textile Industry of the 1970's*, U.S. Dept. of Labor Bulletin 1578 (Washington, D.C.: U.S. Government Printing Office, 1968).

Part III
Strengthening the Analytical Framework

7 Crude Tools and Vulnerable Findings: A Summary Evaluation

Bela Gold

Any summary assessment of the contributions and limitations of the hitherto dominant approaches to analyzing the economic effects of technological innovations must begin by examining the purposes to be served. From these may be derived criteria for evaluating the relevance of the operations and relationships studied, the concepts and measures employed, and the findings and interpretations presented.

The preceding review of the literature in this field emphasizes the haphazardness of the analytical foci dealt with and the underlying vagueness of the purposes of most published studies. And these characteristics seem to be rooted in the implicit, though rarely expressed, fundamental assumption that the important effects of all technological innovations are essentially similar and already well known, except in respect to their specific magnitudes. It would follow logically, therefore, that all empirical probings represent samples from the same barrel, that resulting findings should converge into widely applicable generalizations, and that these would confirm (perhaps also amplifying somewhat) the structure of concepts and theories employed in designing such empirical studies.

As a matter of fact, however, such posited homogeneities have not emerged up to the present. And there seem to be good reasons for doubting that they are likely to emerge soon—or, indeed, ever. Such doubts derive in part from the extremely wide array of innovational developments and operating environments to be encompassed by a generalized concern with the economic effects of technological innovations. Further reinforcement of such doubts derives from the broad range of purposes that may be pursued even within any of the foregoing areas of substantive concern.

Alternative Purposes to be Served

The studies that have been examined reflect a wide spectrum of purposes, which may be combined into a few rough groupings. Most studies have concentrated on measuring the changes in labor, or in labor and capital, inputs relative to total physical output in economic sectors that have introduced technological innovations of some kind. Some have merely reported the descriptive findings represented by changes in average relationships among these variables. Other papers have supplemented such findings with inferences concerning their causes and economic benefits. Even more ambitious undertakings have sought to generalize such inferences as indicative of probable technological effects: (a) in larger sectors of the economy during the period studied; or (b) over future periods within the sector studied; or (c) likely to result from policy changes designed to accentuate or to minimize certain past outcomes.

Systematic coverage of most of these purposes could be expected to strengthen the bases for improved policy formulation and decision-making in industry and government in respect to the rate and direction of technological innovations. But this would require more thorough penetration into the processes whereby technological impacts are transformed into adjustments in resource flows and eventually into changes in economic relationships than is provided by the analytical approaches that have been dominant so far.

This need may be illustrated by examining the potentials and limitations of production functions, the most widely used approach in efforts to determine the economic effects of technological innovations. Production functions are used to predict the relative contributions to changes in physical output of changes in the physical inputs of labor and of capital entering into the output. Application of this approach raises three questions: Who wants to know? What in particular do they want to know for the purposes inciting their interest? To what extent do production function findings fulfill these requirements?

Oddly enough, many economists fail to recognize that managers have no need for production functions at plant or firm levels in the case of single product operations, and regard them as meaningless in the case of multiproduct operations. Such managers have no need to await *ex post* economic analyses to learn the direct effect of changes in labor or capital inputs on product output levels. On the contrary, they know that such relationships are *predetermined* for each product on the basis of engineering and production decisions relating to such factors as product design, material input specifications, the mix of labor skills and manning patterns, equipment characteristics, production schedules, quality standards, and maintenance arrangements. But such knowledge also includes awareness of the loci and extent of differences among products in respect to these numerous determinants of the gross input-output relationships covered by production functions. Accordingly, such average findings for changing product mixes are of no practical value for managerial purposes.

Moreover, product mixes tend to change not only over the long run, but also during business cycles, seasonally, and still more frequently for many industries. Nor can one overlook the tendency for changes in market pressures, in technological potentials, and in input availabilities to engender changes over time in at least some of the above determinants of gross average input-output relationships, even in the case of single product plants. These considerations minimize the value to individual managements of average production function findings over periods of several years.

Of course, such gross input-output relationships are not predetermined at the level of groups of firms or industries—and may not be known at all in the absence of special economic studies. But such ignorance is caused in part by the absence of any management decision-making responsibility at this synthetic level of aggregation.[1] And it results also from the associated difficulties of interpreting resultant findings covering operations whose heterogeneity reflects the effects not only of such interproduct differences as those noted above, but also of interplant and interfirm differences. These latter might include the scale of operations, modernity of facilities, capacity utilization rates, factor prices, financial structure, primary markets, trade union relationships, and other characteristics.

What use could be made of findings covering such heterogeneous operations? By whom? Suppose aggregate physical output rises faster than total man-hours and total capital inputs, what conclusions could be drawn that would be of analytical significance? Certainly, such findings provide no indication of which of the numerous intraproduct, interproduct, interplant, interfirm factors listed above were responsible, and to what extent. Nor does it follow necessarily even that increases in physical output per unit of physical input are a "good thing"—that is, are economically desirable. To make such determinations would require price, cost, and other economic evaluations and might possibly involve certain welfare considerations as well. For example, innovations offering increases in output per man-hour are frequently associated with increases in capital investment and often engender increases in wage rates, thus necessitating further analysis to determine their attractiveness to management.

In short, such production functions do not reveal the causes of observed changes and hence do not indicate either the means by which these changes might be extended (or reversed) or the benefits, if any, resulting from them. Yet these applications represent two of the important purposes of such studies to prospective users of the findings. Of course, after production function parameters have been determined for some composite statistical grouping, vague "explanations" are often tossed in, attributing reported adjustments to "technological progress," "human capital," or other amorphous factors. And these superficial "explanations" are frequently associated with proposals for maintaining or accentuating the trends that have been uncovered—implying the desirability of their effects, although these have not been explored carefully in most cases. Aside from these highly vulnerable interpretations of production function findings, the remaining common application is to build such results into econometric models to be used for forecasting future trends in the industry studied or in larger sectors of the economy. Where these involve essentially statistical extrapolations, their vulnerability is adequately specified by the implicit caveat associated with all such extrapolations: that the expected results depend on the continuation of average adjustment patterns of the past. But how can one estimate the likelihood of such continuing patterns in respect to production functions without analytical insights into the factors responsible for the past relationships?

And the foregoing limitations of production functions at the level of industries tend to grow even more severe at still higher levels of aggregation, such as industry groups, all manufacturing, or the national economy as a whole.

Important decisions must be made at each of the levels of aggregation which have been mentioned. At plant, firm, industry, and national levels, choices must be made among alternative resource commitments. Some must be made by industrial managements, some by financial institutions, some by trade unions and trade associations, and still others by officials of various local and national government agencies. Unfortunately, production function studies have provided few, if any, significant contributions to meeting the specialized needs of these various groups, for the reasons given above as well as for additional reasons relating to measurements, which will be discussed later.

Essentially similar problems limit the applicability of input-output analysis in appraising the effects of technological innovations. Within a single product plant, the technical (physical) coefficients are predetermined by the same technological and production management decisions that were listed above, rather than awaiting discovery via *ex post* economic analysis. Changes in such coefficients are similarly predetermined by engineering and managerial decisions and, hence, their specific causes can be identified, whereas most *ex post* changes recorded by input-output measurements do not permit such specific identification. In respect to multiproduct plants, however, changes in technical coefficients are even less likely to be identifiable through input-output analysis in the common situation involving changes in product mix, whereas these, too, are both predictable and controlled by plant management.

Although most applications of input-output analysis have involved relatively broad economic sectors, obviously every increase in the level of aggregation beyond the single plant entails progressively greater difficulties in analyzing the reasons for whatever results are found. This follows from the fact that cells in an input-output matrix present the gross aggregate value of all materials, all labor and all capital inputs, as well as all product outputs for all plants, firms, and industries encompassed by a defined sector. Hence, as in the case of production functions at comparable levels of aggregation, underlying interproduct, interplant, interfirm, and interindustry heterogeneities prevent determination of the sources of any changes in cell coefficients from one period to another; they also prevent identification of the nature and extent of such heterogeneities within any period.

The basic concepts underlying input-output matrices are enormously instructive, of course, for economists seeking to comprehend and to map resource flows on some intersectoral basis. Moreover, they provide valuable perspectives on the relative magnitudes of component sectors of the economy at given periods. Beyond such descriptive purposes, however, they are much too aggregative to permit effective evaluation of the economic effects of technological innovations except at the level of single product plants. Nor have these, as yet, served any purpose beyond providing unnecessary *ex post* summaries of the effects of specifically identified and managerially approved innovations.

Also in this connection, the preliminary applications of process analysis that have thus far been published seem to suffer from similar limitations. Treating an industry as if it were one integrated plant certainly yields results different from those commonly developed for managerial decision-making and performance evaluations in individual plants and firms with distinctive characteristics. But it is not easy to specify how such average findings covering heterogeneous operations can be interpreted to yield effective determinations of the actual economic results of technological innovations in specific operating environments—whether relating to the input, cost, revenue, and investment effects of concern to industrial managers, or to the employment, market competition, or other effects of concern to government officials.

Alternative Analytical Foci

Alternative analytical foci may be mapped in at least six different dimensions. One of these relates to the kinds and number of innovations to be covered. Very little experience is required to recognize the enormous variety of technological innovations, even if consideration is restricted to those taking the form of hard goods—that is, facilities, equipment, instrumentation and products. These cover a wide range of functions, operating principles, and degrees of specialization; they also differ widely in output capabilities, input requirements, and operating efficiency. Such a crude lumping together of grossly dissimilar developments is akin to aggregating all diseases to determine their average effects. It is conceivable, of course, that this vast multiplicity of specific forms may be classified into some smaller number of categories with relevance for economic analysis. Yet very little has been developed along these lines to date, although this would seem to offer more constructive contributions to the uncovering of subsections of uniformity in effects than dealing with the entire variegated jumble of technological innovations.

In addition to deciding which of the innumerable innovations is to be studied, either empirically or speculatively, the analyst faces the problem of deciding whether to concentrate on single innovations, or on clusters of closely related innovations, or on the unspecified combination of all innovations introduced or operating within a single period. And even when a single innovation is specified as the target of analysis, a critical choice must still be made among the frequently numerous models and their successively improved designs, which usually emerge before a few types eventually establish dominance. Nor can one easily apply the seemingly simple concept of "best practice"[2] as a guide to choosing. What is considered "best" by an array of prospective users differs widely with their specific operating characteristics and with the current relative urgencies of their various technological, labor, marketing, and financial pressures.[3] Thus, there would seem to be no basis for expecting that the results of empirical studies will converge, regardless of which innovations are studied or how many, unless it is assumed that all innovations tend to have the same effects, or unless all effects are ignored except those hypothesized on the basis of assumed homogeneities.

A second dimension of alternative analytical foci concerns the economic effects to be encompassed. These may be so broad as to consider resulting effects only on total unit costs or even on profitability; or they may concentrate on changes in selected major categories of unit costs. Even more specifically, they may focus solely on the changes in input requirements per unit of output that underlie changes within one category of unit costs, such as wages. But even here one encounters important alternatives and associated problems. Technological innovations commonly affect not only the quantity of each input category per unit of output, but also the qualitative characteristics of such inputs and the nature of their productive contributions.

Moreover, technological innovations often involve substitutions among major input categories along with changes in the design and service characteristics of individual products, in the mix of products, and in the level of aggregate output. The spattering of choices among this wide range of alternatives reflected by published studies, theoretical as well as empirical, thus provides another obstacle to fitting their results together—and another basis for doubting the expectation that all innovations will prove to have essentially similar effects.

A third dimension of alternative analytical foci concerns the levels of aggregation of the operating systems to be studied. Some studies have dealt with the effects of innovations in single operating departments, while others have embraced integrated production lines or complete plants. At higher levels of aggregation, studies have covered changes in firms and even in entire industries. Beyond that, various studies have sought to assess the effects of technological innovations for major sectors of the economy and the economy as a whole. It is apparent, however, that higher levels of aggregation encompass increasing analytical difficulties. Even at the level of a plant, the effects of an innovation in any subsector are blended in with continuing relationships in all other operations, thereby tending to yield different results depending on the nature and importance of the given innovation within the given plant. In multiplant firms and in whole industries, even greater variations may be expected, depending on the extent of diffusion of the given innovation, on possible variations in management policies, and on the extent to which other innovations are also being introduced, as well as on the effects of the earlier mentioned differences among products, plants, and firms. Indeed, interindustry comparisons and analyses at the level of national aggregates seem to defy even identification of the innovations whose effects are being assessed, to say nothing of efforts to devise persuasive measures of such distinctive innovational effects.

A fourth dimension of alternative analytical foci concerns the choice made among alternative industrial environments. For example, this might involve concentrating on capital goods or on consumer goods industries, or on industries characterized by large- or small-scale plants. Other significant alternatives might relate to industries differing in the extent of competition in their product or factor markets or differing in the rapidity of technological changes experienced in recent years. And a somewhat different perspective would be provided by differentiating among industries according to the relative importance of governmental as over against market pressures in determining responsiveness to innovational opportunities.

A fifth dimension of alternative analytical foci concerns the length of time to appraise results. It is apparent that the effects of innovations change over time. Short-term changes include modifications in the innovation, improvements in applying it efficiently, adjustments in surrounding operations to reestablish effectively integrated production, and frequently changes in wage rates under trade union pressure. Over somewhat longer periods, other input factor prices are likely to change in accordance with modifications in specifications and with changes in the quantities purchased as the innovation is more widely diffused. Product prices and output levels may also be changed, initially because of the qualitative changes in the product, and later because of the cost adjustments shared by an increasing array of competitors. Moreover, the rates at which such changes are likely to take place clearly differ with the nature of the innovation, the extent of competition in product and factor markets, and consumer responsiveness to changes in product quality and price. Thus, the duration of the period over which technological effects are studied has a major effect on the kinds of influences and outcomes likely to be covered.

A sixth dimension of alternative analytical foci is the specific years covered by the studies made. There is common recognition, of course, of dissimilarities among time periods in respect to changes in general economic conditions such as output levels, prices, employment, capital availability, government regulations, and expenditure policies. Industries and individual firms may also undergo substantial changes from one period to another in their growth, utilization rates, trade union problems, and competitive pressures. Despite enormous and sometimes ingenious efforts, no generally persuasive methods have been developed to exclude the influence of such factors from attempted determinations of the distinctive effects of innovations studied during periods involving such important changes in relevant economic conditions. Hence, the effects associated with given technological innovations in one period may differ materially from the effects experienced in another period of equal duration.

In short, six major dimensions of alternative analytical foci have been outlined for assessing the economic effects of technological innovations, and numerous further possibilities have been identified within each. This broad perspective has several implications. It helps to explain the continuing absence of convergence among published empirical studies. It demonstrates the need to develop a far more complex structure of theoretical expectations along with more penetrating concepts than have guided the formulation of past studies. It also raises doubts about the widespread belief that even the major economic effects of technological innovations are likely to take the form of generalizations which are both widely applicable and also directly relevant to critical evaluative or decision-making issues.

Relevance of the Concepts and Measures Used

Two of the most important components of scientific research involve the progressive reformulation of basic concepts to make them accord more fully with the unfolding characteristics of the phenomena being studied and the development of measures that are increasingly relevant to such concepts. The dominant approaches to determining the economic effects of technological innovations reflected by the published literature leave much to be desired in both respects.

Effects of Technological Innovations

The basic concepts of production functions have a number of significant analytical shortcomings. Perhaps the most important is the attempt to assess the effects of technological changes in purely physical terms. Economics differs from engineering partly in that physical resource flows and relationships must be integrated with their relative prices to permit economic evaluations. Indeed, there can be no economic basis for choosing among alternative technological combinations of inputs and outputs so long as factor prices and product prices are omitted. The production function approach commonly seeks to overcome this obvious difficulty by resorting to even more untenable assumptions; for example, that technological innovations change the quantitative magnitudes of inputs and outputs, but leave essentially unchanged both their qualitative characteristics and their respective prices. But such assumptions eliminate important, and often dominant, effects of most technological innovations.

The emphasis on an essentially physical conception of the relations encompassed by the production function also necessitates resort to dubious measurement expedients. For example, labor inputs are almost invariably measured in man-hours, a concept that reflects an essentially physical volume only so long as changes in the skill mix of such inputs and in the nature of their contributions to the production process are ignored. But both these represent important effects of technological innovations—the former affecting the quality, factor prices, and costs associated with the physical volume of man-hours; and the latter affecting the variability of labor inputs with changes in output levels—for example, the ratio of maintenance and other indirect labor to production labor. The measurement of changes in output in purely physical terms could, of course, be even more misleading. To take economic account of changes in the size and service characteristics of different models of products within even one product line, to say nothing of aggregating such changes in many product lines, obviously necessitates some form of weighting by their relative prices. Efforts to introduce a physical concept of capital inputs are still more dubious because of the extreme heterogeneity of capital goods, the role of capital goods as a prime target of technological innovations, and the resultant absence of effective price comparisons or indexes of average changes in capital goods prices.

Accordingly, the basic conceptualization of production functions as relationships between changes in physical inputs of labor and capital and changes in physical outputs cannot be effectuated at the level of actual measurements in most plants and firms, to say nothing of industries and even larger economic sectors. Instead, efforts are made to approximate this physical concept by deflating the value of output by the change in each product's price to yield the implicit change in total product quantity, and by deflating capital investment (usually after depreciation) by some index of changes in capital goods prices. Such expedients are especially vulnerable, however, when the objective is to determine the economic effects of technological innovations. In part, this is caused by the assumption that the individual products have remained unchanged and, hence, that any attendant changes in their prices were uninfluenced by the technological innovations under study. In the case of capital goods, attendant problems include the inapplicability of any broad index of capital goods prices (the vulnerability of which has already been noted) to a particular sector affected by unique technological innovations, and also the absence of meaningful price comparisons between new capital goods embodying such innovations and preinnovation models. In the case of labor inputs, the inadequacy of man-hours alone is not overcome by the change in average wage rates. This is true because the latter, too, omits the possible changes in labor skills and elasticity attributable to the innovation.[4]

Two additional difficulties are associated with effectuating the concept of physical output in measuring the production of multiproduct operations. When such total output is measured by weighting the quantity of each product by its relative price, technological *advances* that lead to lower unit costs and prices for given products also engender *reductions* in the (price) weights given to such product units in computing total output—whereas *declining* productive efficiency leading to higher costs and prices would tend to *increase* the weights given to such product units in computing total output (Gold, 1971, pp. 203-206). The use of such "economically oriented," or price-weighted, measures of physical output also introduces the frequently overlooked problem of differentiating changes in total output from changes in total inputs. Do changes in product value (that is, in total costs plus profits) at fixed product prices measure changes in total outputs or in total inputs? If the former, as suggested by Fabricant (1939), how are changes in total inputs to be measured? Application of the same approach would suggest measuring them in terms of changes in total costs at fixed factor prices, although differing conceptions of profits might suggest their inclusion or exclusion. But what would changes in such total output to total input ratios signify? Changes in efficiency levels—as implied by terms like "total factor productivity" (Kendrick, 1961) and "aggregative efficiency index" (Schmookler 1952)—or changes in the ratio of (deflated?) total costs to (deflated?) total revenues—that is, some form of (deflated?) profit margins—or changes in the ratio of factor price to product price indexes?

To answer these preceding questions precisely, it is necessary to recall that changes in these ratios may be caused by variations in the price or in the quantity of each specific input per unit of unchanged product, to shifts in input factor proportions, to changes in the product mix of output, and to differential changes between factor and product prices. The resulting amalgam of interacting effects reaches far beyond the implications of any of the simple interpretations suggested above. Nor can it lend itself to any other interpretation, simple or complex, until an analytical framework is developed which is focused on the relationships among such aggregates within an evaluative system having clearly specified objectives.[5]

Still another shortcoming of concept and of measurement arises from the widely prevailing practice in applying the "total factor productivity," as well as the production function, approach of measuring output in terms of value added rather than the value of product, and thereby omitting purchased materials from consideration. This ignores the fact that much of the technological improvement effort within many industries is concerned with changing the kinds and quality levels of materials purchased and altering the level of fabrication at which they are acquired. This is necessary to improve products, to shift to more readily available or cheaper materials, and to eliminate costly or difficult tasks by purchasing inputs already embodying such operations. Thus, the quantity of material inputs per unit of output, and their quality, are as much subject to managerial improvement efforts as inputs of labor and capital—and are just as subject to technological innovations. Moreover, because material inputs frequently account for more than 50 percent of total costs in manufacturing industries, they also affect working capital requirements.[6]

Finally in this same connection, value-added data cannot properly be adjusted by changes in the price of actual final products to calculate "physical" output—inasmuch as material prices and costs need not always parallel those of other inputs or of products, and no data cover the price only of the value added components of the final product. Of course, the inclusion of material inputs in production functions would again pose the problem of deflating material costs, not by some general index of material prices, but rather by deflating each component of material costs by an index of prices related to the particular kinds and quality grades of material directly involved.

The shortcoming discussed above is also rooted partially in another fundamental conceptual inadequacy: regarding changes in physical input-output relationships as separate from, rather than *interactive* with, changes in factor and product prices. This accounts for the prevailing tendency to emphasize changes in such physical productivity ratios as measures of the "real" effects of technological innovations and to attempt such determinations by deflating revenues and costs by accompanying changes in their respective average prices. As has already been noted, however, this implies that technological innovations have had no effect either on the quality or price of inputs and outputs. In reality, of course, changes in the quality and mix of inputs and outputs tend to engender price reactions by increasing the demand for some specific types of inputs at the expense of others and by offering products with altered service characteristics. Moreover, such interactions continue to develop over time as adaptive adjustments to such innovation-induced pressures are made by competitors, by factor markets, and by buyers. This basically untenable separation of the physical relationships from associated economic relationships also encourages the widespread failure to round out such studies by examining the extent to which technological innovations are a *response* to shifts in factor and product prices instead of arising from extraeconomic space and then exerting only a unidirectional impact on resources flows and prices.[7]

From the standpoint of measurement, attempts to encompass such interactions necessitates shifts: in the variables covered, in the duration of the time periods deemed necessary to prevent truncated explorations of effects, and in the form to be taken by reported findings. With respect to the variables covered, requirements would be expanded beyond "physical" inputs of labor and capital (and materials, where these are included—as in input-output analysis) to cover the factor prices of the specific material, labor skill, and capital goods categories displaced as well as those replacing them. Moreover, it would be necessary to measure the rate of diffusion of the innovation under study as well as changes in the level of product output to appraise resulting changes in demand pressures on input factor markets. And attention would also have to be given to changes in the prices of products affected by the innovation With respect to the time periods to be covered, it is evident that the effects attributable to any innovation will unfold more slowly than the diffusion of the innovation itself, because of the frictions involved in spreading information, in resolving conflicting pressures in markets, and in effectuating needed shifts in resource allocations. Hence, the effects of innovations that take ten to fifteen years to account for even one-half of relevant output—as has been common in the steel and related industries (Gold, 1975b, pp. 130-133)—can hardly be determined by studies covering less than twenty or thirty years.

In regard to the form of expected findings, it follows from the foregoing emphasis on interactions over a wide sector of factor and product markets and over an extended period of time, that the common tendency to report *the effects* as net changes in specific variables between some starting date and some evaluation date is likely to be misleading. Instead, results must delineate the *pattern of changes* over extended periods in physical input-output ratios, factor prices, unit costs, and other evaluative criteria deemed relevant by the industrial managers, government officials, or scholars whose questions are to be answered by such studies.

The latter consideration suggests another basic conceptual weakness underlying studies of the economic effects of technological innovations: the pervasive assumption that the dominant effect of technological innovations is to reduce average total unit costs. Of course, this is frequently a primary objective of innovations. But such an objective is too generalized as well as too limited a criterion to guide penetrating analyses. Such undertakings need rather to identify the more specific and immediate objectives of the decision-makers to apply relevant evaluative criteria. For example, their direct objectives might include shifting to more abundant or cheaper materials, or to reduce energy and material waste, or to decrease the rejection rate of products because of quality defects, or to reduce the idle time of facilities, equipment, and so on. But technological innovations may also be adopted for a variety of purposes that may involve increasing rather than decreasing unit costs. The most common relates to innovations yielding new or improved products, which are expected to generate income gains in excess of attendant cost increases. Other instances might involve the adoption of innovations permitting the needed expansion of output through the processing of lower quality inputs, or those required to meet governmental health, safety, and pollution standards.[8]

To take account of such broader objectives, it seems necessary to extend measurements even beyond the structure of physical productivities, factor prices, and unit costs already discussed. In addition, the framework should reach out to include the resulting effects on profitability, if it is to accord with a dominant motivation in the case of industrial adoptions of innovations. And even broader considerations might have to be included to encompass innovations whose adoptions involve governmental incentives, support, or resistance.

The basic point of all the foregoing, of course, is that both the concepts and the measures which have been commonly employed in deriving production functions ignore or offset various significant effects of common technological innovations–thus contributing less to the understanding of such effects than to minimizing them.

Several of these same shortcomings are found in most input-output analyses. Inputs and outputs are measured in terms of aggregate values, ignoring changes in quality and mix. Although the value of material inputs is subdivided according to the totals derived by "an industry" from each of its supplying "industries," labor and capital inputs are usually presented only as aggregates. In addition, all differences among the plants and firms encompassed by a given industry are ignored, including even the extent to which they are utilizing the given innovation or simultaneously applying other technological, managerial, or marketing innovations as well. Hence, this approach, too, prevents determination of most of the specific effects attributable to given technological innovations. Moreover, when comparisons are made between input-output matrices for two periods to appraise intervening changes, the same problems of comparability of coverage and of meaningful deflation discussed above are confronted for all factor and product prices.

Effects of Research and Development Programs

Some attention may also be given to the conceptual and measurement problems involved in studies of the "productivity" and benefits of research and development activities, which are generally regarded as the source of technological innovations. Perhaps the most fundamental weakness of most such undertakings is the implicit assumption that the various activities subsumed under the broad heading of R&D are sufficiently similar among a wide array of firms and industries to warrant analyzing them as if they represented essentially homogeneous operations. Resulting findings of average input-output relationships, average time lags between inputs and benefits, and average rates of profitability seem quite devoid of significance, however, when compared with the extraordinary diversity of the actual undertakings camouflaged by the simple aggregative input and output measures used.

Even simple classifications of all R&D activities into three or four stages fail to convey the far-reaching differences among firms and industries within each such stage in respect to those aspects that are of primary significance for the evaluations being attempted: the specific kinds of achievement being sought, the magnitude of the resources committed, the levels of risk involved, the time perspectives within which final results are expected, and so on. Product improvement targets may range from minor adjustments to meet particular customer needs to providing successive small increments in the attractiveness, operational efficiency, or service attributes of one or more products, or even to creating quite new products. Process improvement targets cover a comparable range and often involve greater difficulties and commitments as well as longer time perspectives than product improvement efforts. In addition, R&D may also include undertakings that: facilitate shifts to lower quality or cheaper material inputs; mechanize labor tasks; improve machine efficiency; meet health, safety, noise, and pollution requirements—and, sometimes, merely service the routine testing of purchased supplies for conformance to specifications.[9]

There is certainly no basis whatever for assuming homogeneity, even among firms in the same industry, in the variety or relative emphasis among such multiple facets of R&D operations. Resulting aggregate measures and average relationships are likely to be no more meaningful than averaging the speeds of a random mix of sprinters, milers, and marathon runners or determining the average characteristics of a random collection of chemical compounds. To avoid such vulnerabilities, it would seem necessary both to develop hypotheses reflecting a more thorough exploration of the dissimilarities encompassed by the R&D label and to test each by means of samples that are more homogeneous in respect to the specified selection criteria.

A second conceptual problem underlying most efforts to assess the returns to R&D efforts derives from the simplistic assumption that changes in the level of payoffs are attributable solely to changes in the levels of R&D inputs. In reality, gains in cost saving, sales, or profits are obviously affected by changes in the effectiveness of all integral components of a firm's operations, including procurement, production, and marketing. Moreover, there is no warrant for assuming either that the contributions of these operations to improvements in the overall performance of a firm are necessarily much smaller and less consistent than those of R&D programs, or that all firms and industries tend to be equally effective in all operations except R&D, leaving the magnitude or efficiency of the latter as the primary source of differences in overall performance. R&D may be a necessary source of contributions, of course, at least from time to time. But emphasizing its role to the neglect of all other operations does not provide a persuasive conception of the complex of internal operations that management must integrate in seeking to improve results.

Also in this connection, calculating the profitability of allocations to R&D by comparing them with net returns at some later date is bound to yield exaggerated results to the extent that this omits all additional investments and costs in other operations, including production and marketing, which proved necessary to convert research findings into new or modified products produced at costs low enough, and sold in quantities and at prices large enough, to yield the eventual profits. For example, a recent editorial in *Science* (Oct. 29, 1976) referred to the "well-known empirical relation in industry . . . for each $1 spent on successful research, $100 must be spent to bring a product to market." For these reasons, few experienced executives are likely to regard the results of statistical studies of the relationship of changes in R&D inputs to changes in sales or profitability as offering sound guides for the allocation of their resources.

There are a number of additional shortcomings in the simplistic conception of changes in a firm's profitability being determined by changes in its R&D inputs. For example, this view tends to ignore the wide range of external factors affecting profitability, including changes in general economic conditions, in the intensity of market competition (as affected by capacity utilization rates, the impacts of imports, the attractiveness of substitute products, and so on), and in the availability and price of needed inputs. It also tends to ignore variations among firms and industries in the extent to which technological innovations may come, not from internal R&D, but from licensing innovations developed by others, or through purchasing equipment and facilities that embody process improvements generated elsewhere. Moreover, the concept of unidirectional causation from R&D inputs to profits hardly accords with the widespread tendency to allocate a specified percentage of sales revenues to R&D, or with the confirmatory complaints by industrial research directors about sharp budget cutbacks during periods of declining sales and profits.

A third conceptual problem confronting efforts to assess the benefits of R&D involves the assumption that the contributions of R&D can be measured in the form of reduced costs or increased revenues and profits. But this ignores the fact that such efforts seek to achieve a competitive advantage concurrently with similar efforts by competitors. Accordingly, it seems necessary to consider several alternative possibilities. If all competitors made equal commitments to R&D, had the same targets, and were equally successful on average, sometimes one and sometimes another would reach the goal first. On average, however, no firm might have a consistent competitive advantage, even if all competitors kept increasing the level of R&D inputs. Under less restrictive assumptions, it is still reasonable to argue that a substantial part of any firm's R&D would have to be "defensive," in the sense of trying to minimize either the advantages currently achieved by the R&D efforts of one competitor or another, or the disadvantages resulting from differentially unfavorable adjustments in its own input factor prices or supplies. In either of these cases, even successful R&D efforts would have the effect merely of reducing comparative cost, sales, and profit disadvantages instead of yielding positive gains in these areas.[10]

More generally, it is puzzling to find regression results being presented and interpreted without accompanying efforts to present a persuasive line of reasoning which presumably underlies the hypothesis being tested. For example, studies of whether R&D intensity increases with the size of the firm, the degree of concentration, or the extent of diversification frequently fail to offer supporting arguments that could not be reversed without much loss of cogency. Such analytical underpinnings might be strengthened by considering how specified differences among firms might impinge on each of the intermediate stages linking R&D inputs with payoffs. For example, the very concept of industrial research implies seeking to advance sufficiently beyond the current knowledge and practices of competitors to yield commercial benefits. Such efforts are obviously attended by risks, whose magnitudes tend to grow rapidly as targets become progressively more ambitious. Hence, differences in this first stage, involving the ambitiousness of R&D objectives, may be the major source of differences in failure rates and in the time lags between R&D inputs and payoffs—and may be a more important source of differences among industries, and perhaps between large and small firms, than later stages in the development of R&D efforts into sales and profits.

In the next stage, involving the conversion of R&D outputs into commercial levels of production, one might suppose that efforts to develop product improvements would engender smaller risks and shorter delays as well as more limited additional investments than major process innovations. This would identify another source of differences in R&D profitability that, however, might well cut across the firm and industry categories most closely associated with observed differences in the first stage. And in the third stage of this rough classification, the effectiveness with which new processes or products are marketed surely represents another important source of differences in R&D profitability—perhaps more closely associated with the size and financial resources of firms than with various other characteristics. Similarly, if one were concerned with the effects of the relative diversification of firms on the profitability of their R&D efforts, it would seem helpful to clarify both the expected advantages and disadvantages of diversification in this connection. The latter might include the necessity of dispersing R&D resources, as well as production and marketing improvement efforts, over a wider array of products and processes. Advantages might include the availability of a wider range of processes and products to harness the often diverse outputs of R&D efforts.

Effects of Diffusion

Most research on the diffusion of technological innovations in industry has measured the rates of diffusion for particular innovations and sought to identify the factors associated with observed differences in such rates. The key conceptual problems for both center around seeking to define "the innovation" and its practical range of application meaningfully, and they derive from the very dynamics of technological improvement processes.

As has been widely explored by others (see, for example, Merton, 1973), major technological advances frequently appear in several forms. They then tend to grow in numbers as increasing efforts are devoted to improving their effectiveness and to modifying them to fit the specialized needs of major groups of prospective users. In the case of facilities and equipment innovations, resulting changes are likely to affect operating efficiency and reliability, scale of through-put, input specifications, product quality, and product range, as well as labor skills and safety. Contrary to the suppositions of outsiders, such developments are often difficult, costly, and time-consuming. And yet diffusion rates may depend on the realization of improvements of these kinds.

As a result, if these stages of metamorphosis are ignored on the basis of hindsight, one may well wonder why innovations that eventually became dominant took so long to gain wide acceptance. On the other hand, recognition of the fact that each stage of development had only limited diffusion potentials would require more detailed field studies than have yet been made of the advantages and limitations of each significant improvement as well as of the resulting new application opportunities. One need hardly add that such undertakings would have to be far more penetrating than those which have yet been published, and would probably alter most, if not all, published findings to date.

Similar conceptual limitations have characterized analyses of the factors affecting the diffusion rates that have been measured. The overwhelming emphasis has been on expected cost savings and profitability, although *ex post* analysis has generally failed to demonstrate that such expectations were fulfilled, much less that these were in fact the determining motivations for adoption of the innovations studied. Of course, adoption decisions must have been made because of expected benefits. But it does not follow that these would have to take the form of demonstrable decreases in costs or increases in profits. Other powerful benefits, which were mentioned earlier, might have included mitigating expected increases in input factor prices or decreases in the availability of needed materials; responding to buyer demands for higher quality products; acceding to governmental pressures relating to labor or community health, safety, and pollution standards. Nor can it be assumed that the expected benefits of given innovations are so clear that all competitors would assess them similarly or even that all competitors give serious consideration to the same innovations in any given period. But these doubts suggest that the criteria applied to the evaluation of available innovations may differ widely among firms, reflecting differences in their internal urgencies, resource availabilities and specialized expertise rather than deriving solely from the demonstrable benefits of the innovation itself. Hence, the foci of efforts to explain differences in diffusion rates may have to be expanded to include consideration of a variety of firm characteristics (Gold, 1975a, pp. 142-46).

Curiously enough, little serious research has been done on the advantages and disadvantages of slower *versus* faster rates of diffusion, or even on the nature, patterns and effects of the diffusion of industrial technologies. It has been a commonplace for economists to complain about the slowness with which various major innovations have been utilized.[11] But such criticisms have invariably been based on hindsight judgments with little detailed consideration of the options, pressures, and uncertainties confronting prospective adopters during successively earlier stages in evaluating the available forms of the innovation.[12] Instead of assuming ignorance, sloth, bias, or stupidity as the causes of such restrained rates of diffusion, it would be more helpful to make field studies of the actual considerations and evaluations responsible for the decisions made.

Also worth emphasizing in this connection is that maximizing diffusion rates might prove undesirable for economic as well as social reasons, because many innovations prove technically unfeasible or uneconomic except in specialized conditions, and because many others become attractive only after extensive technical improvements or changes in economic conditions. Hence, over eagerness in adopting innovations can involve extended periods of high cost operations, customer dissatisfaction, and even heavy capital losses. Indeed, although a number of well-known cases of pioneering adoptions have yielded large rewards—usually relating to products rather than processes—there is still no consistent evidence favoring early adoption of all innovations or indicating how to separate those which are still subject to serious risks of obsolescence or unprofitability from those already warranting adoption. Surely the development of deeper insights into the time patterns of risks and rewards associated with successively earlier and later adoptions of actual technological innovations would represent more constructive contributions than continued criticisms on the basis of only superficial analysis combined with hindsight.

Both understanding and evaluation of technological diffusion requires fuller exploration of resulting adoption patterns and their effects. The usual focus of diffusion studies is limited to competitors in the given industry. But the diffusion of technological innovations may also have important effects on the factor markets supplying the industry, including materials, labor, capital goods producers, and even suppliers of investment funds. In another direction, such diffusion is likely to have repercussions on the markets of the products directly affected as well as of competitive substitutes. Additional effects may concern government agencies, including effects on competition, on natural resource requirements, on employment and labor relations, on foreign trade and on the environment. Only in this larger context can the effects of different diffusion rates be adequately evaluated as a basis for clarifying the possible advantages and disadvantages of seeking to alter them.

Vulnerability of Interpretations of Past Findings

Serious studies of the economic effects of technological innovations must begin by recognizing that such innovations engender readjustments in economic relationships: (a) through a variety of intermediate linkages; (b) within a network of interacting pressures; (c) covering an expanding horizon of operations; (d) over extended periods of time; and (e) accompanied by a variety of independent but concurrent developments. But production functions, total factor productivity determinations, and input-output matrices do not provide such detailed mappings of the means through which technological innovations bring about eventual economic changes. Their reliance, instead, on inferring causal factors from changes in the gross relationships between inputs and outputs is not only a superficial expedient, but lends itself to serious misinterpretations.

For example, as was noted earlier, the exponential values of labor and capital inputs in production functions commonly are interpreted as indicating the extent to which an increase of 1 percent in man-hours, or of 1 percent in capital inputs, would increase total physical output.[13] This interpretation rests, however, on the assumption that each factor is paid the value of its marginal product, although the theoretical conditions that are necessary to fulfill this assumption are seldom satisfied in reality. Indeed, close analysis of the means whereby these variables are measured—usually involving attempts to deflate value added, wages, and some measure of capital charges by more or less relevant price indexes—offers at least comparable support for a quite different interpretation of these exponential values. Specifically, they might be interpreted as representing merely the average proportions in which value added has been divided between labor and capital payments. And such income (or cost) shares need not imply any comparable proportioning either of their respective actual contributions to increasing output or of their relative physical productivities.

Misinterpretations of results are also frequent in studies of the effects of increases in mechanization. These usually report increases in output relative to labor inputs and decreases (or smaller increases) in output relative to capital inputs, leading to interpretations that the innovation has increased labor productivity and decreased capital productivity (or increased it less than labor productivity). In fact, closer analysis suggests that, in most such instances, parts of labor's productive tasks have been shifted to machinery, thus reducing labor's contribution to output along with the reduction in the volume of labor inputs—leaving indeterminate whether labor's productive contributions per man-hour have risen, remained unchanged, or declined. But trade unions are often able to secure increases in wage rates in proportion to increases in output per man-hour, even when achieved via mechanization (as is commonly the case). Hence, deducing changes in labor's contributions from changes in its wage rates by applying purely theoretical assumptions, and without more penetrating analysis of the means whereby the technological innovations effectuated the observed economic changes, encourages quite serious misinterpretations. On the other hand, closer analysis of the actual technologies involved, and of the processes whereby innovations bring about changes in resource qualities and flows, would permit more perceptive identification of changes in the relative contributions of various input factors, as well as in other characteristics of the operations affected. As has been shown elsewhere, clarification of the constraints on factor substitutions in various technologies also facilitates more effective solutions to the classical problems of determining the relative contributions of different inputs to joint products (Gold, 1955, pp. 215-30).

Another illustration of the vulnerability of speculative interpretations of changes in the gross levels of relationships provided by production functions relates to the effects over time of a cost-saving innovation affecting one or several of a plant's products. During the period of early adoptions, the total cost of producing the same level of physical output of the affected products would decline, thus yielding the expected decrease in average unit costs for the plant. But suppose that progressive diffusion combined with competition among adopters of the innovation leads to price reductions for these products to take advantage of producers still relying on the costlier processes. Because changes in total physical output are calculated by weighting the quantity of each product by its average price over the periods being compared (if the Edgeworth formula is used), the reduction in the affected products' relative prices has the effect, as was noted earlier, of reducing their contributions to the total plant's "physical output" as reflected by price-weighted measures of product quantities. Hence, if total costs remain the same in this second period, but the accompanying measure of the plant's physical output declines (not because of the number of product units, but because of the price decreases effected by the innovations), then total costs relative to physical output rise! In short, factor savings that lead to cost savings and then to price reductions are accompanied by initial reductions in unit costs, which then tend to erode. Accordingly, evaluations over periods long enough for competition to have forced such a pass-through of production cost savings to consumers would tend to minimize the unit cost changes that had taken place—even though the cost and price per physical unit of output had actually declined and remained lower.

Shortcomings in respect to interpreting the results of studies of the economic effects of technological innovations are not solely attributable to ignorance of the complex of interactions connecting the initiating physical resource impacts of innovations with various adjustments in economic relationships. No less disappointing, from the standpoint of the decision-making and policy-formulating concerns of industrial and government officials, is the pervasive failure to improve our understanding of the sequential linkages in transmitting the various impacts of different kinds of innovations; the sources of frictions and even of offsetting pressures in these successive stages; and the timing and duration of the major stages. Insights of this kind would strengthen the bases for adoption-or-rejection decisions by permitting consideration of probable long-term as well as short-term outcomes. They would also facilitate setting realistic performance goals during the period of integrating the innovation into the larger complex of operations. And attention would be drawn to the need for detecting and trying to overcome possible developments that could undermine expected benefits. So long as reliance continues to be placed on the gross aggregative relationships that have been the foci of most production function, total factor productivity and input-output studies, there is reasonable basis for questioning whether the reported economic effects of technological innovations were even largely, much less wholly, attributable to the innovations from which they supposedly derived.

Such reported results are also subject to still another source of misinterpretation. This stems from the widespread failure to recognize the gross unrepresentativeness of the experiences covered by most of the published literature. Perhaps the single most important reflection of sampling bias is the overwhelming concentration on those innovations which hindsight reveals to have been substantial successes. It is quite conceivable, however, that these account for only a minuscule proportion of R&D and other innovative undertakings and expenditures—as is suggested by the meager number of such triumphs reported by Mueller for one of the leading industrial research groups in the United States over a thirty-year period marked by heavily increasing outlays for research.[14] In view of this possibility, much more discriminating studies would seem to be called for in order to provide more realistic perspectives on the proportions of all innovational undertakings or expenditures which result in commercially attractive outcomes. And such studies would be even more valuable to industry and government if they were to disaggregate findings according to various categories of innovational efforts.

Other aspects of unrepresentativeness in the reports that have been summarized include overconcentration on a few industries (especially pharmaceuticals, chemicals, petroleum refining, and steel) and on relatively large firms with reputations for being unusually progressive. It should be emphasized that there is nothing objectionable to exploring new fields of analysis by taking advantage of the most readily available points of entry. Difficulties develop only when such fragmentary probings are used as the basis for unjustifiably broad generalizations with policy implications.

Some critical attention might also be given to the data employed in these studies. As Kuznets (1950) pointed out long ago, governmental statistical data tend to be biased in accordance with the purposes for which they are collected. Hence, in an environment where R&D expenditures are regarded as an evidence of progressive management and where low ratios of such outlays are regularly criticized, it would not be surprising to find statistics on R&D expenditures tending to be inflated by widespread resort to broad conceptions of the varieties of activities to be covered. Along somewhat similar lines, experiences with mailed questionnaires has demonstrated that those with favorable experiences to report are more likely to reply than those with unfavorable experiences. Even in an interview situation, respondents are more likely to provide answers that reflect favorably on themselves and their organizations than the opposite, and tend to be reluctant to admit ignorance concerning matters about which they might reasonably be expected to know. Research must, of course, be carried on despite such common difficulties. But awareness of such potential biases emphasizes the need for extended probings, cross-checks and critical evaluations before essaying generalizations with far-reaching implications.

Conclusions and Implications

It is understandable, of course, that new problems in every field tend to be attacked at first with the concepts and tools aready in hand, even though they were developed to deal with problems that had emerged in the past. Only increasing evidences of the inadequacy of such available weapons, combined with increasing pressures for more effective progress, are likely to overcome the reluctance to abandon familiar approaches in favor of groping along unaccustomed paths and bearing the attendant greater probabilities of frustration. Production functions, total factor productivity, and input-output techniques were not developed to determine the economic effects of new technologies. As such determinations became increasingly important, it was natural to use the available tools to gain at least preliminary insights into this newly burgeoning problem area. And such expedients undoubtedly provided useful initial perspectives on some of the relationships involved. But there seems to be little doubt of the need to strike out in new directions if economists are to contribute more helpful guidance in the development of policies relating to the rate and direction of technological innovations.

Studies of the economic effects of technological innovations are undertaken for a variety of purposes. Among the most urgent are the specialized needs of business executives who make decisions to invest or not to invest in developing and applying technological improvements. Equally pressing are the distinctive needs of government officials who are concerned both with assessing the national and more localized problems and opportunities engendered by technological advances and with formulating policies to encourage such developments as would maximize resulting benefits to the nation. In addition, as was noted in Chapter 1, economists face similarly broad arrays of analytical urgencies, including the need to develop more penetrating insights into the factors affecting the level of resources allocated to promoting technological advances, shifts in the targets of such efforts, variations in the rates of application of emerging advances, the widely varying responses to such advances in factor and product markets, and the speed with which the effects of innovations reach a peak and are then dissipated.

None of these sets of needs gains substantial help from studies that are overgeneralized in their coverage of technological changes or that involve analyses at levels of aggregation far removed from the impacts directly relevant to the issues confronting business and government decision-makers. Thus, studies of the purported economic effects of the combination of all technological changes lumped together are of little value for such purposes because technological changes obviously differ widely in the magnitudes, patterns, and duration of their effects as well as in their sources, nature, gestation periods, and diffusion rates. Similar limitations affect the applicability of studies focused on broad industry groupings whose component firms differ widely in respect to their specific technologies, product mixes, input requirements, scales of operation, and competitive pressures. Nor are practical decision-making needs helped significantly by studies that correlate R&D and other innovational investments with growth or profits several or many years later, without effectively differentiating the effects either of concurrent nontechnological company and market developments or of the other efforts and costs usually involved in overcoming intervening production and marketing (and sometimes procurement and labor) problems.

The applicability of such studies also tends to be limited by the extent to which findings have been shaped by assumptions and methodologies that insulate them from decision-making environments and evaluative criteria. For example, businessmen are not likely to exhibit great interest in determinations of what the results of technological innovations "would have been" if neither the qualities nor the prices of input factors and products had been changed since the innovations were introduced. Their experience has amply demonstrated that most major product and process innovations tend to engender interactions with factor and product prices through altering the specific types and quality grades of inputs as well as the quantities required, thereby affecting the relationship of demand to supply for the categories affected and through altering the product qualities that affect the prices and levels of demand which they can command. At any rate, they have a greater concern with the results likely to appear on their financial statements than with what would have happened if product and factor markets had been frozen.

Decision-makers find it difficult to interpret the significance of production functions that compare changes in *price-weighted quantitites* of products with *physical* measures of man-hours, *net investment* measures of capital, and, at times, *cost* measures of material inputs. Nor is it easy for them to accept the assumption that input factors are paid in proportion to their marginal productivities, even in this imperfectly competitive world, and hence that the exponents in production functions really reflect the relative physical contributions to output of the input factors covered, rather than merely their shares of total revenues (or value added). And there is ample basis, as has been noted, for uneasiness about the seemingly incidental manipulations involved in deflating revenues, costs, and investments to estimate changes in physical inputs and outputs.

Reported results are also difficult to interpret because of wide differences among them and because of the patent unrepresentativeness of the samples on which they are based. Hence, although there is widespread agreement that successful technological advances are desirable, little is known about their frequency, their net returns to developers and pioneering adopters relative to attendant costs and burdens, or the patterns and duration of their broader economic and social effects. In short, prevailing findings reflect the inevitable superficiality of analyses from afar, reliance on crude concepts and tools, and substituting imagination and logic for close and detailed exploration of the phenomena to be understood.

Accordingly, most of the generalizations about the effects of technological innovations and increasing allocations to R&D programs that have gained widespread acceptance are based on the shaky foundations of very few and unrepresentative samples, crude tools, weak measures, and dubious interpretations. This would be of little significance if only peripheral matters were involved. But R&D and technological innovations are among the most important sources of growth and competitive strength, not only at the level of firms and industries but even at the level of entire national economies. And it is precisely because major issues of resource allocations and of related policy development are involved that it seems unduly hazardous to be guided by comforting, but essentially untested and highly vulnerable, generalizations.

In view of the increasing opportunities offered and threats posed by competitive efforts to generate technological advantages, it seems urgent to intensify efforts to build stronger research foundations for discriminating between excessive allocations in one industry and inadequate allocations in another; between setting goals that are too high or too low; between accelerating diffusion rates and slowing them down; between emphasizing basic research as over against commercial development; between priorities favoring product improvements and those favoring process improvements. Resulting research strategies accordingly would tend to pass over the currently popular, but meaningless, issue of whether total R&D expenditures are sufficient or whether they should be expanded. Instead, such strategic perspectives would tend to emerge in the form of portfolios of gradually changing allocations among various intermediate categories of R&D and innovational undertakings, between the extremes of "low risk-short duration-small reward" projects and "high risk-long duration-large reward" projects.

Different industries might then be expected to develop portfolios with dissimilar allocation profiles, and disparities of this kind might also appear among firms in the same industry—as has often proved to be the case in product mix, sales, procurement, and investment strategies. What is far more important than the extent of parallelism among such technological improvement strategies or allocation portfolios is to develop more effective experiential and analytical guides to developing such portfolios in accordance with the objectives and resource availabilities as well as the current and prospective competitive pressures of particular firms, industries, or nations.

The resulting need for deeper probing of the chains of successive effects produced by technological innovations carries with it several important implications. First, the simple and fragmentary analytical approaches that have been reviewed must be replaced by a more comprehensive framework providing systematic coverage of the network of variables involved in the processes of interaction between technological and economic factors.[15] Second, the application of such a complex framework to a wide variety of technological innovations in different industrial settings and under disparate economic conditions is likely to narrow the range of direct applicability of resulting findings. Moreover, the economist's traditional drive for maximum generality of results is also likely to be undermined as changing technologies curtail the time span within which valid evaluative comparisons can be made. And a third and perhaps the most serious implication of such a more comprehensive approach is the necessity of acquiring considerable highly specialized knowledge about the industrial sectors studied to comprehend the rich texture of technological changes involved and to trace their successive interactions with physical input and product characteristics, and with the factor and product markets as well as with the price, cost, investment, and other economic aspects of the operations affected.[16]

Economists have been free to avoid such burdens so long as they were willing to concentrate on general scholarly issues. But increasing efforts to provide more effective guides for public and industrial policies require more thorough knowledge of the forces, processes, and pragmatic criteria encountered at the levels of aggregation and in the particular sectors of the economy where decisions are made and where their effects are lived with. As a result, analytical models are likely to become more complex and more numerous, and economists—like scientists, engineers, and medical doctors—may have to develop and even admit to narrower areas of expertise than the current self-image of the "general economist."

Notes

1. For a discussion of important changes in the foci of decision-making at different levels of aggregation, and in attendant evaluative criteria, see Gold (1971, pp. 44-48).
2. As an illustration of this theoretical emphasis on "best practice," which then proves inapplicable in the following empirical sections of his analysis, see Salter (1960).
3. For a more detailed review of such pressures, see Gold (1975a, pp. 142-44).
4. For an extended discussion of these problems, see Gold (1955, pp. 15-45).
5. For further discussion, see Gold (1971, pp. 56-58).
6. The cost of purchased materials and supplies have averaged about 50 percent of total value of products for "all manufacturing" in the United States for more than 75 years, according to the Census of Manufactures. For examples, see Gold (1971, p. 137) and (1975a, p. 12).
7. For a detailed analysis along these lines, see Gold (1971, pp. 193-97).
8. For a more detailed discussion, see Gold (1975a, pp. 142-44).
9. For a more detailed discussion, see Gold (1971, pp. 222-36)
10. For a more detailed discussion, see Gold (1975b, pp. 24-25)
11. For example, see Adams and Dirlam (1966) and Ault (1973).

12. For example, see Dilley and McBride (1967) and Huettner (1974).

13. For example, see Samuelson (1976, p. 538).

14. See Mueller (1962, pp. 323-58).

15. For an example of such a framework, see Chapter 8.

16. A widely accepted contrasting view is expressed by Solow (1965) as follows: "The economist really need not know at all what it feels like to be inside a steel plant—he quantifies technological change by making measurements of output per hour, or output per unit of this or input per unit of that." For a detailed discussion, see Gold (1971, pp. 32-59).

References

Adams, W., and J.B. Dirlam, "Big Steel, Invention and Innovation," *Quarterly Journal of Economics* (May 1966).

Ault, D., "The Continued Deterioration of the Competitive Ability of the U.S. Steel Industry: The Development of Continuous Casting," *Western Economic Journal* (March 1973).

Dilley, D.R., and D.L. McBride, "Oxygen Steelmaking—Fact vs. Folklore," *Iron and Steel Engineer* (October 1967).

Fabricant, Solomon, *The Output of Manufacturing Industries, 1899-1937* (New York: National Bureau of Economic Research, 1939).

Gold, Bela, *Foundations of Productivity Analysis* (Pittsburgh: University of Pittsburgh Press, 1955).

_______, *Explorations in Managerial Economics: Productivity Costs, Technology and Growth* (London: Macmillan; New York: Basic Books, 1971).

_______(ed.), *Technological Change: Economics, Management and Environment* (Oxford: Pergamon Press, 1975a).

_______, "Alternative Strategies for Advancing a Company's Technology," *Research Management* (July 1975b).

Huettner, D.A., "The Development of Continuous Casting in the U.S. Steel Industry: Comment," *Economic Inquiry* (June 1974).

Kendrick, J.W., *Productivity Trends in the U.S.* (Princeton, N.J.: Princeton University Press, 1961).

Kuznets, S.S., "Conditions of Statistical Research," *Journal of the American Statistical Association* (March 1950).

Merton, R.K., *The Sociology of Science: Theoretical and Empirical Investigations* (Chicago: University of Chicago Press, 1973).

Mueller, W.F., "The Origins of the Basic Inventions Underlying DuPont's Major Product and Process Innovations, 1920-1950" in *The Rate and Direction of Inventive Activity: Economic and Social Factors*, ed. R.R. Nelson (Princeton: Princeton University Press, 1962).

Salter, W.E.G., *Productivity and Technical Change* (Cambridge: Cambridge University Press, 1960).

Samuelson, P.A., *Economics*, 10th ed. (New York: McGraw-Hill, 1976).

Schmookler, J., "The Changing Efficiency of the American Economy," *Review of Economics and Statistics* (August 1952).

Solow, R.M., "Education and Economic Productivity," in S. Harris and A. Levensohn (eds.), *Education and Public Policy* (Berkeley, Cal.: McCutcheon, 1965).

8 Proposed Conceptual and Analytical Improvements

Bela Gold

To overcome the continuing shortcomings of widely used methods for determining the economic effects of technological innovations that have been discussed, it seems necessary to probe more deeply into several elements:

1. The distinctive characteristics of various types of technological innovations, instead of assuming them to be essentially homogeneous from the standpoint of economic analysis
2. The succession of intermediate effects by means of which the initial physical impacts of such innovations engender an array of changes in economic relationships within the firm and beyond
3. The extent to which innovational effects may be shaped by both managerial preferences and external pressures.

Revised Concepts of Technological Changes and Their Effects

To help sharpen the following proposals, I will concentrate on those elements of technology which directly affected the functioning of industries. Moreover, despite the obvious importance of advances in knowledge, skills, and organizational arrangements, my primary focus will be on the categories of technological change manifested in the hard goods involved in industrial operations. Among these, the most important will probably be among the following:

1. *Modifications in the physical and chemical properties of material inputs, as well as the introduction of new types* (a) to offset shortages or decreases in the quality, or increases in the price, of customary materials; (b) to minimize transportation costs; (c) to reduce waste in processing; and (d) to improve product attributes.
2. *Changes in the design and scale of operating processes, facilities, and equipment* (a) to improve speed, precision, yields relative to inputs, safety, and reliability; (b) to increase capacity; (c) to permit the production of new or improved products; (d) to permit adaptation to changed material inputs; and (e) to reduce the level and to alter the composition of labor requirements.
3. *Improvements in control systems* (a) to increase automaticity; (b) to mini-

mize deviations from desired operating conditions, flow rates, and product quality; and (c) to facilitate effective adjustments to changes in output levels and product mix.

4. *Alterations in the design of products* (a) to improve the quality of services to users (including convenience, flexibility, durability, reliability and safety); (b) to provide new kinds of services to users; and (c) to reduce the input requirements and ease the production problems of manufacturers.

Such categories have proved useful in efforts to organize the helter-skelter outpouring of technological change for analytical purposes. However, many of these developments—especially important ones—cannot be confined within any one category. The fundamental reasons are the technological and economic pressures for the progressively tighter integration of successive stages of operation to conserve energy and time, to reduce intermediate handling and inventories, and to promote the mutual reinforcement of adjacent control sectors. Thus, the creation of new kinds of production facilities frequently entails modifications in the kinds of inputs best adapted to them as well as alterations in production control devices and in the characteristics of resulting products. When such accompanying changes are substantial and recognized at the outset—which happened when open hearth steelmaking furnaces were introduced, for example—all the above categories are affected simultaneously by what is usually referred to as one major innovation. But even substantial developments may initially be concentrated within one stage of operation (or one of the above categories), only to exert gradually mounting pressure for adaptive adjustments in successively earlier as well as successively later stages of production, as may be illustrated by the basic oxygen furnace. In the latter instance, analysts may be led to report several innovations, each of which readily fits into but one of the categories of technological change.

Effective appraisal of the economic consequences of such a variety of technological innovations seems to require revisions of the prevailing analytical framework. Most past studies have yielded what may well prove to be seriously misleading conclusions concerning the magnitude and effects of technological innovations because they relied on measures of inputs, outputs, unit costs, and prices that assume that technology has changed only the quantities of input factors and products, excluding the possibility of significant alterations in their distinguishing characteristics as well. The resulting empirical findings of stable production functions, neutral technical change, constant returns to scale, and the proportioning of embodied versus disembodied technical change derive in large measure from the crudities of this unidimensional view of a multifaceted development process.

The wide array of industrial operating characteristics that may be affected by technological innovations is indicated by the following illustrations *merely at the physical level:*

1. *Changes in physical inputs* involving not only the average quantity of each factor per unit of output, but also the kinds, quality grades, and uniformity of the various purchased materials used; the specific skills required and the composition of relative skill grades in labor inputs as well as attendant restrictions on work loads; and such aspects of capital facilities and equipment as productive capacity, degree of specialization, normal working life, flexibility of use levels, and prospective obsolescence rates.

2. *Changes in physical output* involving not only its aggregate volume, but also the variety of product lines, the array of quality grades and sizes in each, and alterations in their service features for buyers, including capacity, precision, durability, reliability, safety, and convenience.

3. *Changes in physical aspects of production flows*, including the controllability of output rates and the speed with which they can be altered; the variability of different inputs with changes in output levels and product mix; the average duration of the production cycle; allowances for maintenance and other down-time; inventory requirements; and associated accident, health, and environmental hazards.

Only by penetrating to such detailed levels of operating requirements and outcomes can technological innovations be evaluated without ignoring resulting chains of interacting adjustments within the production complex that determine their eventual effects on output requirements, factor prices, cost functions, working conditions, prices, and demand as well as profitability.

Revised Analytical Framework

To meet such practical requirements of decision-making in industry and policy development in government, it is necessary to encompass a far more complex structure of relationship than is currently available. It is also desirable, in the interests of maximizing coverage and relevance while minimizing additional costly data collection, that such a framework should utilize as fully as possible the records already generally available in industry as the by-product of existing managerial control and evaluation processes, including procurement, production control, sales, and accounting data.

Measuring the Physical Flow Effects

With respect to measuring the physical effects of technological innovations, the above needs would seem to require coverage of:

1. Changes in the nature and level of each category of input requirements per unit of ouput, including materials, facilities investment, and salaried personnel as well as direct labor

2. Changes in the proportions in which inputs are combined, both to take account of substitutions (such as buying more highly fabricated components instead of making them or replacing labor with machinery) and also to differentiate between changes in the unit input requirements of major as over against minor factors of production
3. Differences between the unit input requirements of factors when they are fully utilized and when their contributions are reduced by idleness (such as underutilized equipment)
4. Simultaneous variations in all components of this "network of productivity relationships" to provide guidance for managers capable of adjusting relationships among any of them in the interest of improving aggregate performance relative to specified criteria.

One means of meeting these needs is offered by the following analytical model of relationships among direct inputs (Figure 8-1). It identifies six components of the "network of productivity relationships," including the "apparent" productivities of labor, materials, and fixed investment as well as the proportions in which these direct inputs are actively combined with one another. Each may be the initial point of impact of technological innovations subject to managerial control. But a change in any component requires adaptive adjustments in other components of the integrated process, subject to both market constraints and constraints imposed by the nature of industry's technology.

For example, the partial displacement of labor by new machinery tends to increase the ratio of actively utilized fixed investment to man-hours and, hence,

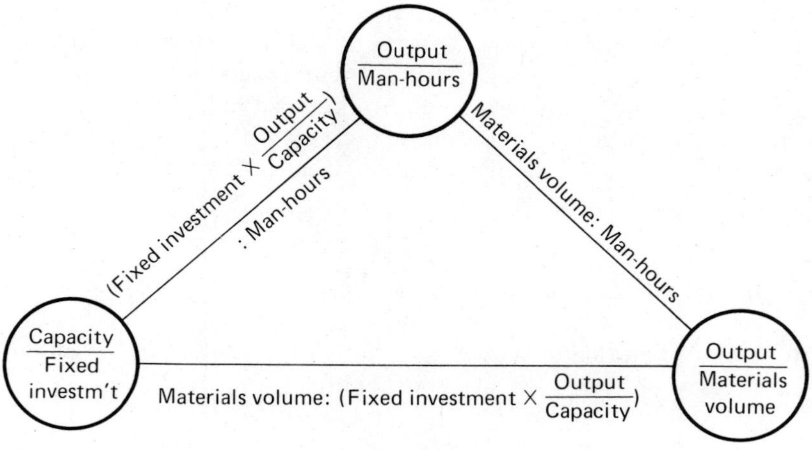

Figure 8-1. The Network of Productivity Relationships Among Direct Input Factors

output per man-hour. In turn, this would require either increasing materials inputs, if output is raised in proportion to the gain in output per man-hour, or decreasing labor input if output remains unchanged. Consideration must then be given to whether the originating innovation involves any change in the nature as well as the volume of unit material requirements and also to any accompanying changes in the productivity of net fixed investment. Technological constraints differ according to whether the processes involved are "labor-dominated," "capital-dominated," or "material-dominated," as will be discussed later.

Measuring the Cost Effects

Whatever may be the technological desiderata of engineers, however, the appraisal of technological innovations from the economic point of view—which is critical for managerial purposes—cannot be restricted to predominantly physical input-output relationships. Specifically, management cannot evaluate the net benefits of a past innovation solely on the basis of data specifying resulting adjustments in each of the six components of the network of productivity relationships. Nor can management choose between alternative innovations on the basis solely of estimated effects on each of these six components. The analysis must be extended to include appraisal of their economic implications if it is to be a sound basis for managerial decisions. And a first step in this direction is to explore the cost effects of changes in unit input requirements and factor proportions by superimposing the "structure of costs" onto the "network of productivity relationships," as shown in Figure 8-2.

The effect of changes in output per man-hour on unit wage costs depends, of course, on concomitant changes in wage rates. Similarly, the effect of changes in unit material requirements on unit material costs depends on accompanying changes in the price of such materials. And the effect of changes in the productivity of fixed investment on the cost of each investment per unit of output depends on the annual rate of charges on such investment as well as the rate of utilization of capacity. What is being emphasized in this framework, however, is the necessity of considering *interactions* between productivity adjustments and factor prices instead of continuing to make the simplifying assumptions that the latter remain unchanged.

In turn, the effect of a change in unit wage costs on total unit costs depends on the proportion of total costs accounted for by wages as well as concomitant changes in other unit costs weighted by their respective shares of total costs. For example, if wages account for less than one-fifth of total costs—as is common in United States manufacturing—a 5 percent decrease in unit wage costs would tend to reduce total unit costs by only 1 percent. But total unit costs are more likely to increase than decrease if the innovation engendering the decline in unit wage costs involved increases in unit material and unit capital costs, which together

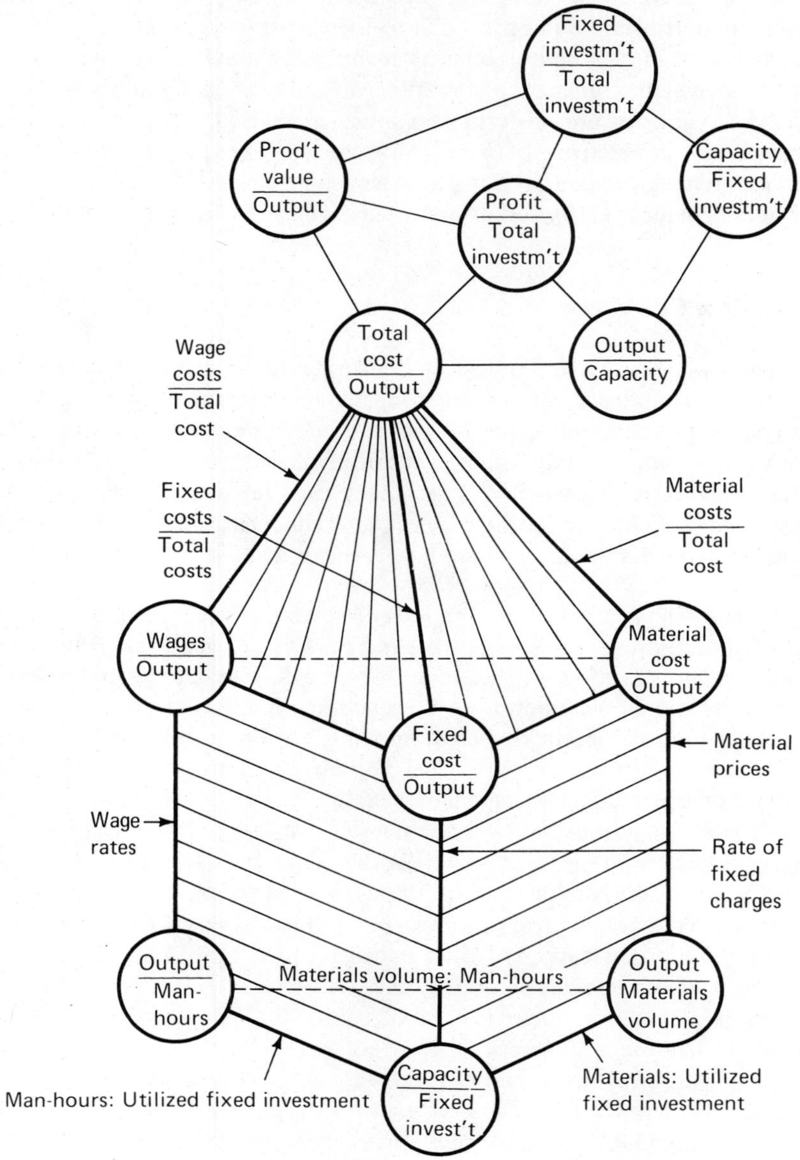

Source: Reprinted with permission from *Omega, International Journal of Management Science* Vol. I, February 1973. © Pergamon Press, 1973.

Figure 8-2. Productivity Network, Cost Structure, and Managerial Control Ratios.

account for more than three times the wage share of total costs in many manufacturing industries.[1]

Integration of the productivity network and cost structure models thus relates changes in "apparent" input productivities and factor proportions through factor prices to each of the unit costs; it also relates changes in individual unit costs through cost proportions to total unit costs. Hence, it identifies the additional kinds of information required to evaluate the prospective effects on total unit costs of given patterns of changes in the network of productivity relationships engendered by technological innovations.

One point in the preceding illustration may well warrant renewed emphasis: that analysis of the effects of productivity adjustments must begin by identifying their sources—whether these are technological innovations or other developments. This is necessary to prevent overlooking other concomitant adjustments, while concentrating on those which are most controversial. Thus, labor-management disputes about the extent of increases in output per man-hour and about the proper magnitude of attendant adjustments in wage rates often divert attention from the means (often technological) and costs of effecting the changes in production processes that account for all or most of the changes recorded in output per man-hour.

Covering Broader Economic Effects Within the Firm

However, managerial decisions in private industry obviously cannot be based on minimizing total unit costs either, in view of the overriding importance of the rate of profit on investment. Accordingly, to evaluate the actual or prospective effects of technological innovations, the productivity network and the structure of costs must be further integrated with some model of the determinants of changes in profitability. One, the "managerial control ratios" model, suggests that variations in the rate of profit on total investment are traceable to changes in and interactions among average product prices; total unit costs; capacity utilization; the productivity of fixed investment; and the internal allocation of investment between fixed and working capital:

$$\frac{\text{Profit}}{\text{Total Invest.}} \equiv \left(\frac{\text{Profit}}{\text{Output}}\right)\left(\frac{\text{Output}}{\text{Investment}}\right)$$

$$\equiv \left(\frac{\text{Product Value}}{\text{Output}} - \frac{\text{Total Cost}}{\text{Output}}\right)\left(\frac{\text{Output}}{\text{Capacity}}\right)\left(\frac{\text{Capacity}}{\text{Fixed Invest.}}\right)\left(\frac{\text{Fixed Invest.}}{\text{Total Invest.}}\right)$$

According to this system of "managerial control ratios,"[2] it would follow that efforts to increase the rate of profits on total investment need not involve concentration on cost-reducing innovations alone. Indeed, the latter would be deemed undesirable if achieved by means leading to reductions in product prices and capacity utilization levels that more than offset the prospective gains in profitability derived from unit cost reductions. Conversely, innovations might be deemed attractive if they offer prospects of increases in capacity utilization and in product prices great enough to offset any increases in total unit costs, whether caused by higher priced input factors or heavier input requirements per unit of output.

Figure 8-2 shows how these "managerial control ratios" may be integrated with the network of "productivity relationships" and with the structure of "cost relationships" to provide a unified framework for systematically exploring the complex of interactions linking changes in factor inputs and factor prices to unit costs and cost proportions and to the other determinants of changes in the rate of profits on investment. This framework, which may be referred to as the PCP (or Productivity-Cost-Profit) Network Approach, can be used in analyzing past innovations, in developing innovational strategies for achieving specified future targets, or in appraising alternative innovations, even when their initial impacts focus on different parts of the system.

Intrafirm Applications of the Analytical Framework

One of the major problems of analysis cited in preceding chapters is the use of measurements that are too aggregative to delineate the distinctive effects of particular technological innovations, thus yielding results that combine such effects with those of various other unspecified intrafirm and extrafirm developments. To overcome this difficulty requires development of an analytical framework that permits systematic and progressively deeper disaggregation until the direct impacts of any given innovation come into the central focus of analysis—and then permits broadening the focus to trace the successive ripples of effects emanating from the initial points of impact. It is important to emphasize, therefore, that all the framework presented in Figure 8-2, except the superstructure of managerial control ratios, can also be applied to the component operating units within plants, including departments and even cost centers within departments. Moreover, such cost center results can be aggregated through successive hierarchical layers.

Specifically, as a first step, changes in total unit costs for the plant can be traced back to changes in the total unit costs of each department (or cost center), provided that care is taken to include only incremental rather than total material costs in sequential operations, lest earlier material inputs be counted repeatedly. As a second step, changes in the total unit costs of each operating unit can be traced back to changes in each of the major categories of unit costs weighted by its proportion of total unit costs, as suggested by Figure 8-2. Continuing with this analytical framework, changes in the major categories of unit costs would be traced back to changes in the average factor prices and in the average quantities of inputs per unit of output for each of these major categories.

Although the foregoing subdivisions provide deeper penetration than the dominant approaches discussed in preceding chapters, it still does not get down to the specific resource choices that are directly involved in most technological innovations. For effective evaluation of such past or prospective innovations, it is usually necessary to subdivide materials, labor, and capital input categories into the various specific types that they encompass and that are the direct targets of managerial efforts both to increase the efficiency with which each is used and to make substitutions among them. The process of elaborating the productivity network for such deeper penetration is illustrated in Figure 8-3, which represents the ironmaking department within a steel mill. Each of the three circles at the bottom level of Figure 8-2 is now decomposed into its more homogeneous components. For example, output/materials volume is first separated into fuels, iron-bearing materials, and other materials—and these are then subdivided once again into the specific types that are actually dealt with. Thus, fuels include coke, oxygen, and oil and gas; but the latter might be divided further if management were concerned, for example, with the need to consider shifts among fuel oil, natural gas, and internally generated by-product gases. Similarly, iron-bearing inputs offer the possibility of changes in the efficiency

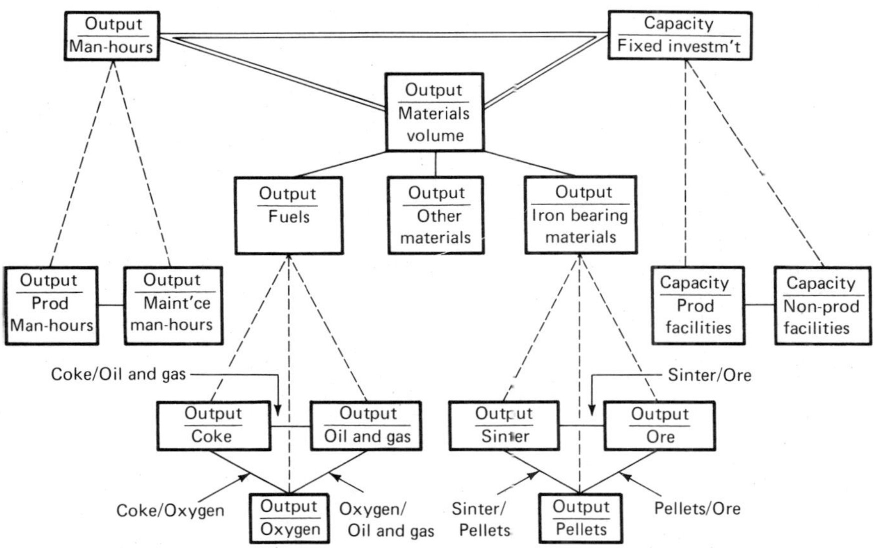

Figure 8-3. Elaborated Productivity Network for Blast Furnaces

with which natural ores, sinter, and pellets are utilized and also shifts in their relative proportions, in addition to changes among specific types of natural ores and pellets. It should be emphasized that the lines connecting the inputs at any level of decomposition represent their proportions to one another, as in Figures 8-1 and 8-2, and as is partially illustrated for fuels and iron-bearing materials in Figure 8-3.

Effective utilization of this detailed network of physical productivity relationships, however, obviously requires parallel elaboration of the superstructure connecting it to changes in unit costs, as illustrated in Figure 8-4. Thus, instead of tracing changes in total unit material costs, for example, to changes in the total volume of all heterogeneous materials per unit of output—an essentially statistical concept devoid of managerial or analytical significance—and to changes in the average price of this meaningless composite, the pattern of decomposition is shifted. Changes in total unit material costs would instead be traced first to changes in the somewhat more homogeneous and more meaningful categories of fuel costs per unit, iron-bearing material costs per unit, and, for example, other material costs per unit. In turn, changes in each of these would be traced to changes in the unit cost of each of its still more homogeneous components. Thus, changes in total iron-bearing material costs would be traced to changes in natural ore costs per unit, pellet costs per unit, and sinter costs per unit. And only then would changes in each of these, for example, pellet costs per unit, be traced to changes in the quantity of this much more homogeneous input per unit of output and to changes in its much more meaningful measure of price.

What purposes are served by such detail? To begin with, it should be obvious that although the possibilities of elaboration are quite extensive, only such factors of the analytical framework have to be elaborated and applied as are relevant to the evaluations being undertaken. If these concern the effects of a process innovation in one department, for example, the analysis may be restricted to the inputs, outputs, and related costs of that department. If, however, the innovation has also altered the quantity, quality, or mix of outputs, these further effects must be traced through as many subsequent operations as reflect such carry-over effects to a significant degree. In the same way, any effects on the quantity, quality, or costs of inputs to the department adopting the innovation must be traced through successive antecedent operations. Conversely, it seems obvious that the distinctive effects of specific innovations cannot be determined through measurements at a level of aggregation where such effects are blended in with a variety of concomitant but not necessarily related developments.

Even more important for managerial purposes, it is only through such detailed tracing of successive pressures emanating from the innovation that it is possible to identify possible alternative adjustments as the basis for choosing those which are most advantageous. Moreover, only a comprehensive knowledge of these adjustments, ramifications and attendant costs and delays makes it

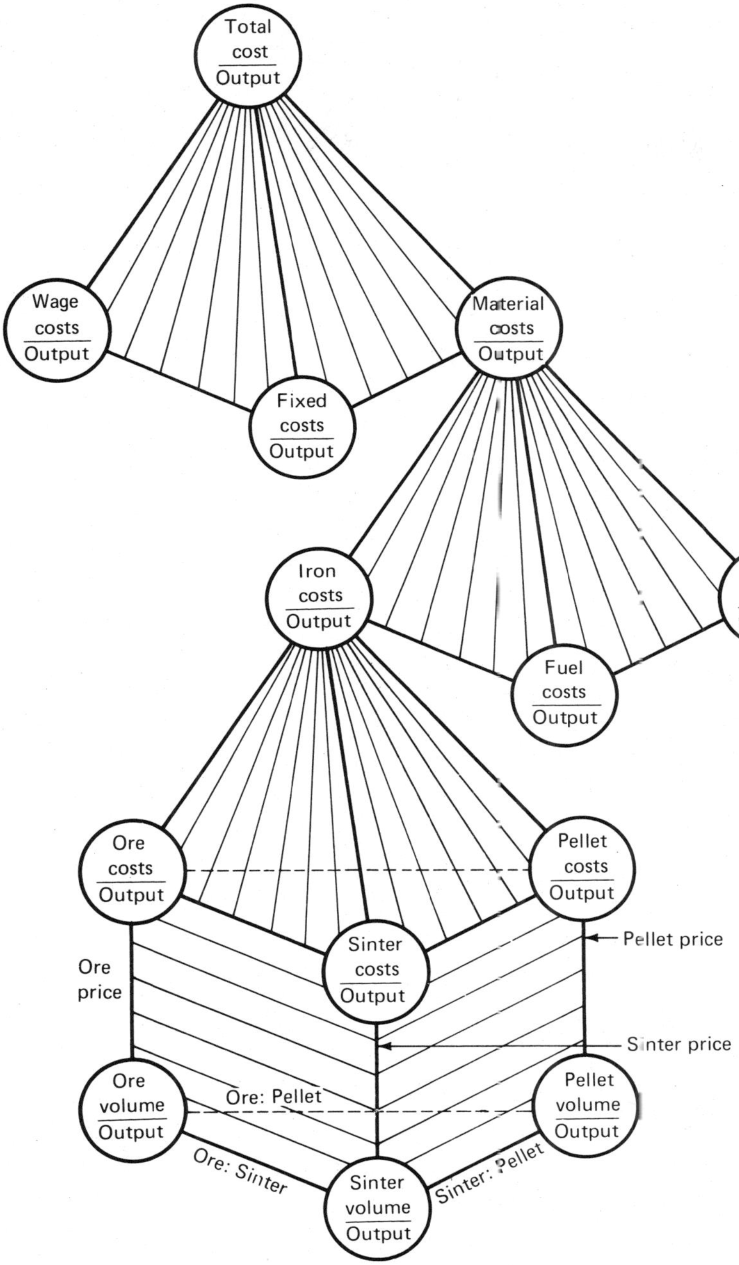

Figure 8-4. Illustrative Detailed Productivity Network and Cost Structure for Intraplant Analysis

possible to evaluate the net effects of the innovation realistically. And a fuller grasp of the range, patterns and timing of such successive pressures and effects would make it possible for managements to develop more effective estimates of the probable outcomes of prospective new innovations.[3]

Improving Measurements

Attempts to apply the preceding analytical framework to empirical determinations of the effects of technological innovations are confronted by a number of problems concerning the specific measurements to be made as well as the coherence of the evaluative comparisons used to assess outcomes. One set of these problems arises from the already mentioned tendency of technological innovations to alter the qualitative dimensions and mix as well as the sheer quantities of inputs and outputs. This raises questions not only about how to encompass all three, but also about the evaluation of accompanying factor and product price adjustments. A second set of problems centers around the need to assure congruence in the coverage of the measures which can properly be juxtaposed for evaluative purposes, as in the case of productivity ratios and unit cost changes.

Output Levels and Product Prices

Changes in the output of multiproduct operations between two periods have long been calculated by economists, as was noted in Chapter 7, by weighting the quantities of each product in each period by its average price in the two periods and then dividing the resulting total for the second period by the total for the base period (that is, by using the Edgeworth Index or some equivalent). Resulting measures of change reflect what has been called "an economically oriented concept of physical output" or a "cost absorption concept of physical output" (Eilon, Gold, and Soesan, 1976, pp. 92-97). This is certainly superior to crude physical concepts for aggregating different products that are still encountered occasionally, such as total tons of steel products shipped or the total number of automobiles produced. But such price-weighted indexes of aggregate qualities are nevertheless subject to certain weaknesses that are especially important when used in analyses concerned with the effects of technological innovations.

Specifically, such measures tend to ignore changes in the qualitative characteristics of products, both by including only their quantities and also by treating any accompanying changes in product prices as though they were attributable solely to inflationary or other market factors unrelated to technological changes—thereby also ignoring the reciprocal effect on such prices of technology-induced changes in the quality of products or in production processes. In addition, the use of final product prices as weights in measuring changes in output levels involves the assumption that there has been no change in the (cost) proportion of the finished products representing materials and supplies purchased from external sources. Moreover, the use of price weights, which are clearly relevant to business evaluations, obviously need not be correlated with the changing service values provided by technologically improved products to consumers.

To help reduce the effects of such shortcomings, certain simple expedients may be employed. When a product has undergone substantial changes in service attributes, it should be entered into the Edgeworth formula as a new product rather than as continuation of the old product series. Hence, it is included in the output of the second period, weighted by its current price alone, whereas its predecessor is included in the output of the first period, weighted only by its then-current price. When products have undergone more limited, but nevertheless significant improvements, their resulting price increases may be expected to exceed the average price increases of essentially unchanged products—and this differential may be used to adjust the actual quantities of such products in the second period to show that each product unit represents a somewhat larger volume of output than formerly from the standpoint of an economically oriented concept of physical output. Similarly, when technological innovations have altered the ratio of purchased materials cost to the value of all products—either decreasing it because of backward integration or increasing it through buying more highly processed supplies than formerly—the calculated change in physical output between the two periods need only be adjusted inversely with the change in this ratio.

The remaining problem of input measurement arises from its multidimensional nature. Thus, the cost absorption approach obviously yields different results from those obtained by using the technical criteria of engineers, or the weight and volume criteria of shippers, or the service criteria of consumers. For example, when the average mileage per tire was doubled by the addition of some chemicals to the rubber, consumers could count each tire as two in terms of use-expectancy, but production managers could not do the same in terms either of costs absorbed or revenues generated. Hence, the eventual reduction in product prices in accordance with technologically induced reductions in production costs—as a result of the diffusion of the technology and of competition among its adopters—produces a decrease in the economically oriented concept of such output. This tends, as noted in Chapter 7, to erode the reductions in total unit cost recorded prior to the decline in prices. But such economic readjustments in product and factor markets need have no effect on the enhanced service values of the products. In short, even though the cost savings and related profits of those adopting the innovation gradually diminish and even disappear, the attendant service benefits to users do not. And no evaluation of the effects of technological innovations could be regarded as comprehensive that omitted recognition of such social benefits.

Input Levels and Factor Prices

In empirical studies, measuring inputs involves even more problems than measuring outputs because there are usually many more categories to deal with, because these are far more dissimilar, and because each is subject to variations in more dimensions that are relevant to explorations of the economic effects of technological innovations. And these difficulties are intensified by the obvious need for present purposes to reverse the hitherto dominant thrust of analytical efforts toward aggregative results.

The multiplicity of inputs is apparent from the fact that even medium-sized plants may purchase hundreds of different materials, components, and supplies, may employ labor in dozens of skill categories, and may utilize hundreds of different kinds of machinery and equipment. As for measuring such inputs, obvious options include measuring what is paid for, or what is actually used, or what productive contributions are provided. Labor inputs, for example, might be measured in terms of numbers employed, total wages and fringe benefits paid, man-hours paid for, man-hours worked, skills applied, energy expended, and hazards borne. Capital inputs might be measured in terms of total assets, total fixed investment (net of depreciation), number or volume of facilities and equipment, available productive capacity, productive capacity used, and capital charges. And purchased supplies cover an even broader range of commodity categories and of physical, chemical and other qualitative differences within each—along with changes in quantities purchased, quantities delivered, quantities used, and attendant costs and inventory charges. Finally and obviously, inputs and outputs change in nature, quantity, and value with each successive stage in production processes.

How are choices to be made among this large army of alternatives? The only reasonable answer would seem to be that they would have to vary with the specific objectives of the evaluations to be made. Accordingly, the level of aggregation of the measures might be focused, as suggested earlier, on the cost centers or departments of primary impact and then proceed to antecedent and subsequent operations before moving up to plant and firm results. In turn, within each of these foci, the sequence of analysis would follow the structure presented in Figure 8-2 and, for deeper penetration, would shift below the level of total costs to the more detailed framework illustrated in Figure 8-4.

This approach clarifies some of the choices to be made among alternative measures of inputs by illustrating the use to be made of cost measures at one stage and of physical measures at another, and also by illustrating the separation of heterogeneous aggregates into progressively more homogeneous components, depending on the level of analytical concern. The operational criteria for appraising homogeneity usually center around function and price. For example, the two primary subdivisions of labor inputs may well be by major skill categories (that reflect significant differences in wage rates), and between directly productive labor as over against maintenance and other indirect labor—which obviously affects the flexibility of wage costs. Meaningful distinctions would also be made between fuels used to produce heating and lighting, for example, as opposed to fuels entering into processing operations. Distinctions would also be made among grades of a given raw material, such as coal, ore, or petroleum, characterized by substantially different prices and offering quite different levels of contributions to operations.

Capital, Inputs, Capacity and Utilization

In measuring capital inputs, it is usually helpful to separate investment in fixed assets from working capital and to separate investments in land and buildings from investments in productive facilities and equipment. The latter category of inputs can then be measured in financial terms by the investment in such capital goods net of accumulated depreciation; and in physical terms it can best be measured by its productive capacity. This expedient may be defended on the pragmatic grounds that managers determine their capital goods requirements in accordance with the planned nature and volume of productive capacity and only as a derivative follow-through do technical specialists translate this guiding decision into specific numbers of various types of equipment and facilities. Thus, instead of confronting the problems of somehow combining the productive capacity of each of perhaps hundreds of machines and other capital goods, most of which perform highly specialized operations quite unlike those done by others, it would seem thoroughly realistic to measure the productive capacity of the entire installation.

In measuring capital inputs for comparison with other inputs and with outputs, three additional considerations should be taken into account.[4] First, there need not be any consistent relationship over periods of a few years or less between capacity levels and net fixed investment. This is true partly because depreciation allowances tend to reduce net fixed investment more rapidly than accompanying declines in capacity, especially during the first half of expected equipment life. But it is also caused in considerable part by the fact that most plants contain bottlenecks at some stages of production which prevent full utilization of productive capacities in other stages. Hence, limited expenditures on easing some bottlenecks can lead to disproportionate increases in through-put capacity. As a result, it could be seriously misleading to interpret increments or decrements in net fixed investment as having engendered accompanying increments or decrements in capacity. Sounder insights are likely to be offered by comparing total capacity with total net fixed investment and then interpreting any changes in the relationship with a clear understanding of depreciation patterns in mind.

A second aspect of measuring capital inputs involves the need to recognize the varying and frequently substantial gaps between the stock of such inputs on hand and the proportion being actively utilized for productive purposes at any given time. Both measures are important along with the differences between them. As a practical matter, such determinations can be made by calculating productive capacity, then measuring output and using the ratio of output to capacity as a measure of the utilization rate. Accordingly, net fixed investment can be multiplied by the rate of utilization to measure the "actively utilized" portion of net fixed investment, which is included in the factor proportions linkages of Figure 8-1 and later figures.

Incidentally, the concept of utilization ratios is increasingly becoming relevant to measures of labor inputs as well. Gaps between man-hours paid for and man-hours at work are attributable to paid vacations and holidays, sick leave, wash-up time, coffee breaks, and, in some industries, to time for internal transportation within the plant or mine or construction site. Gaps between man-hours at work and man-hours effectively utilized in production are largely attributable to the decreasing flexibility of man-hours relative to variations in production levels, whether because of work rules, increases in the ratio of indirect workers, or high levels of guaranteed wage payments.

Returning to the measurement of capital inputs in the form of available productive capacity, it should be emphasized that, although measuring the latter poses extremely difficult problems within the purview of pure theory, a practical estimate is usually readily available in most plants because of its fundamental role in production planning, scheduling, and control. Moreover, it is usually available in precisely the form of direct relevance for present purpose—that is, as an array of estimates covering each common pattern of product mixes, because the capacity of any multiproduct plant obviously varies with the mix of products to be produced. And this practical knowledge is incorporated into the production planning that guides the scheduling of machines, the routing of orders through different operations, and the specification of delivery dates. To calculate total plant capacity exactly in accordance with output measures, thereby yielding proper capacity utilization rates, it is only necessary to calculate capacity by the same techniques as output. Specifically, for each product mix, the capacity level of output for each product would be weighted by its current price and the resulting aggregate would be compared with that resulting from weighting the actual output of each product by the same current price—yielding a percentage utilization rate for exactly congruent product mixes.[5]

Productivity Changes

Merely juxtaposing the comparative magnitudes of specified inputs and outputs reveals nothing more than the level of, or changes in, the given ratio. Evaluating such findings requires that the variables be derived from an analytical framework which encompasses all the inputs and outputs of the system and provides a theory of how it functions. Such a framework permits working backward from specified performance objectives to determine which key variables should be studied, how they should be related to one another, and what measures of each would maximize the usefulness of findings. Where such theories are vague—such as the determinants of optimal functioning in marketing, education, or research—productivity measurements cannot be designed which are meaningful for evaluating the performance of such systems.

The significance of given input-output ratios depends not only on the analytical relevance of the categories used, but on five additional requirements reflecting the forms whereby technological innovations insinuate themselves into aspects of operations with economic and managerial import. One concerns the qualitative stability of input and output categories through time, lest changes in their nature should confuse interpretations of observed adjustments in given input-output ratios. For example, ostensible changes in the productivity of smelters may actually be traceable to variations in the metal content of the ores fed into them. A second warns of the dangers of quantifying peripheral rather than core aspects of input and output flows, as may be illustrated by the previously mentioned use of tonnage shipped to measure the output of steel mills, although most production efforts beyond the furnaces seek to *increase* the value of products by *changing* the shape, and incidentally *reducing* the weight, of the steel being processed. The three remaining requirements are that the numerator and denominator of given productivity ratios should relate to congruent sectors of activity; that they should relate to properly linked time periods; and that the contribution of the input must be absorbed into, and affect the characteristics of, the output. These may be illustrated briefly by noting that the total input of a plant should not be compared with only part of its output (for example, relating total man-hours of integrated steel mills to the ingot tonnage output of the furnaces); that in operations involving extended production cycles, output levels should be compared not with concurrent materials input, but rather with those earlier inputs represented in the current outputs; and that input measurements must encompass all categories which may be substituted for one another (thus it would be easy to misinterpret changes in the ratio of pig iron input to open-hearth steel output if no account were taken of the use of scrap in place of pig iron) (Gold, 1971, p. 32).

Unit Costs and Real Costs

The original conception of unit cost as the sum of the outlays involved in producing a discrete unit of product is clearly inapplicable in modern large-scale manufacturing operations, which usually involve the simultaneous processing and fabrication of many streams of components that are eventually assembled into a variety of products. Under such conditions, unit costs really represent the average ratio of total costs to total output over a given period rather than an average of the actual costs of each unit of output. At the same time, the increasing importance of indirect inputs—such as planning, control, product development, training, and supervision—further complicates the concept of unit costs because of uncertainties concerning the incidence of their contributions as among products and time periods.

But an even more troublesome issue concerns what is meant by the average unit of product which represents the denominator of unit costs. If such costs have been calculated as noted above, the output of plants producing a variety of types, sizes, and models of several product lines would, presumably, be expressed in terms of the statistical measure representing an implied "unit of composite product," which is patently nonexistent. Suppose, as a simple example, that a plant producing refrigerators of various sizes and with different combinations of features found that the average production cost per refrigerator had risen by 6 or 7 percent in the preceding year, while the composite average unit had changed from 7.5 to 8.75 in cubic feet of volume, along with some changes in motor design, insulation, freezer space, and defrosting arrangements. Clearly, the interpretation of such findings presents continuing difficulties because every change in the "composite product" would undermine unit cost comparisons from period to period as the basis for evaluating plant-wide performance, and because the concept of a composite product is meaningless from the standpoint of production as well as sales.

To help interpret changes in given unit costs (or product prices) over time, these are often deflated (divided) by accompanying changes in the general price level as a means of estimating changes in "real" costs (or prices), that is, in the quantities of resources absorbed into the products involved. But this technique, too, reflects inadequate recognition of the pervasive effects of technological innovations on the economy. Thus, resource-saving innovations tend to affect factor prices, to spread from their points of origin to other firms and industries using such inputs, and to engender changes in the unit input levels and factor prices of substitute resources. In turn, changes in the general price level represent an average of resulting price changes experienced by a large array of commodities. Deflating the total unit costs of any plant or industry by changes in the general price level can be interpreted most effectively, therefore, as indicating the extent to which the given unit costs have changed as compared with accompanying changes in the average cost of resources absorbed (plus profit) per unit of output in the rest of the economy.

It follows that a horizontal trend in deflated unit costs, as reported for various industries (see, for example, Gold, 1971, pp. 183-193, 203-207), does not signify an absence of technological progress or an absence of benefits from such progress—as would be implied by the interpretation emphasizing no change in "real" costs—but simply that such unit costs have moved in unison with the average price of all products. Such a relationship could mean that economies in the plant (or industry) have roughly paralleled those in the economy; or that the industry has withheld the benefits of its differentially greater economies from consumers to be shared by its input factors through higher wage rates, salary levels, or profit rates; or that these input factors have received relatively lower payments to absorb the burdens of the industry's differentially smaller economies as compared with other industries. Indeed, the deflated total unit costs of any industry could deviate from a horizontal trend over extended periods only if there were a progressively expanding margin (either upward or downward) between its factor prices and profits and those in other industries—a condition likely to be undermined over time by the mobility of resources and by competition from substitute products.

Two comments may prevent misinterpretation of this section. First, because inadequacies of data and limitations of tools seem to be inevitable problems in empirical research, it would be even more inexcusable to hold up such work until entirely satisfactory tools and data become available than it would be to ignore the shortcomings of existing, makeshift tools and data. Nevertheless, a critical review of this literature can hardly avoid the judgment that a substantial proportion of empirical studies fail to minimize the likelihood of misinterpretations of their findings and conclusions both by avoiding the temptation to overgeneralize and by reminding readers of the limitations of the tools and data employed. Second, identifying past shortcomings does not imply that they could have been avoided in the past; nor does proposing future improvements imply that all are immediately realizable in all situations.

Changing Evaluative Criteria with Increasing Aggregation

The relative importance of the various physical, financial, and welfare attributes of input and output flows tends to change with the level of aggregation. Within centrally controlled systems (such as firms), evaluative criteria tend to be derived downward from the guiding objectives of the whole through successively smaller subdivisions; controls over resource flows tend to be developed backward from the intended output pattern through successively antecedent stages to initial inputs. Hence, all input-output adjustments must be interpreted within the context of their specified restrictions.

Within factory components producing one part for a marketable product, physical input-physical output relationships tend to dominate performance evaluation. Output can be measured only in terms of the number of parts meeting the required specifications. Although inputs can be measured in either physical quantities or costs, the former tend to dominate performance evaluations within such component operations because the qualitative specifications for labor, materials, and equipment, as well as their prices and charges, are usually set by other units of the organization (such as procurement, engineering, industrial relations, and accounting). Welfare constraints are also specified for the given unit by health and safety standards, insurance protection and other components of company policies, trade union agreements, and social legislation. Thus, physical criteria are dominant only because financial and welfare conditions are fixed outside the given unit. Indeed, significant changes in any of these prescribed constraints would make it difficult to interpret resultant changes in physical input-output ratios without careful evaluation of their effects within a larger framework.

In evaluating the larger system that manufactures the complete product, however, cost supersedes physical quantity as the aspect of input flows to be compared with physical output within any given set of physical and welfare constraints. This reflects the production manager's freedom to improve performance not only by decreasing the quantity of each input per unit of output, but also by altering the qualitative specifications of inputs, adjusting factor proportions, and seeking reduced factor prices—all derived from his emphasis on lowering total unit costs. But interpretating unit cost adjustments presents difficulties if accompanied by substantial changes in product design, product mix, or other constraints—as discussed in the preceding section.

At the level of the firm, the primary criteria for performance evaluation changes again. Within defined physical and welfare constraints, reducing costs relative to physical output becomes less important than increasing revenue relative to costs; and increasing profit relative to total investment becomes still more important. Thus, one may note a shift in performance measures (from physical input-physical output) to (financial input-physical output) to (financial input-financial output) at these successively higher levels of aggregation within the firm—as shown in Figure 8-2. This does not, of course, deny the relevance of "physical input-physical output" measures of performance. On the contrary, it reinforces the earlier point: most changes in aggregates are the passive result of changes that actually occur at the level of component operations; therefore, effective managerial control of the system also requires an integrated hierarchy of differentiated measures reaching downward from aggregate "financial input-financial output" measures through intermediate "financial input-physical output" measures to "physical input-physical output" measures, both to trace the sources of changes in the aggregate financial ratios and to plan and implement the readjustments in the underlying relationships necessary for achieving desired improvements in the aggregate financial ratios.

In short, because the variety of physical, financial, and welfare aspects of input and of output flows may interact with one another, changes in any given aspect cannot readily be interpreted unless other significant aspects remain fixed. Moreover, the relative importance of these different aspects changes with the focus of the evaluator. And one must also recognize that, although improvements in the input-output ratios of one sector may generate benefits at more aggregative levels, these may also be offset by (and even cause) unfavorable adjustments in other components of the larger system. Such considerations emphasize the dangers of measuring selected input-output relationships without defining relevant constraints, of applying identical measures to different levels of aggregation, and of projecting observed measures from one level to another.

And an additional array of differentiated analytical needs are encountered at levels of aggregation beyond the firm, as will be discussed later.

Constraints on Harnessing the Benefits of Technological Innovations

The benefits derived from any form of input-output adjustments can take the form of different patterns subject to managerial choice. Two groups of factors tend to narrow the range of such alternative allocations. The first, primarily changes in output and in productive capacity levels, tends to engender similar limitations in all industries. The second, derived from basic technological characteristics, tends to produce differing limitations among industries.

For present purposes, it may suffice to illustrate the influence of the first group of factors with two simple examples. Thus, an increase in output per man-hour (or in output relative to material inputs) could lead to changes in the total requirement for the input factor ranging from reductions to increases depending on managerially determined adjustments in total output. But that output decision may also affect the total input requirements of other inputs—tending to alter total materials inputs in proportion to the change in output and tending to affect capital input requirements in varying degrees (or not at all) depending on the rate of capacity utilization prior to the planned change in output. And such changes in total requirements for various inputs may engender responsive adjustments in factor prices over time—especially as the initiating innovation and associated output adjustment diffuses more widely among competitors.

Unlike the factors just discussed, which tend to exert similar pressures in all industries, management options are dissimilar among industries whose technologies impose disparate restrictions on the proportions in which input factors can be combined effectively. In particular, the most significant aspect of such technologies in this connection is the comparative roles of direct labor, direct materials, and capital facilities and equipment in determining productive capacity.

One extreme, which might be characterized as "capital-dominated," includes processes in which most production workers start, stop, load, unload, set controls, or otherwise simply facilitate the functioning of machines and other capital facilities.[6] In such processes, such as electricity generating plants, petroleum refineries, and steel mills, materials would play merely the passive role of providing the standardized inputs being acted upon by the machine process. In such capital-dominated processes, changes in the productivity of fixed investment (that is, productive capacity-fixed investment) are attributable primarily to changes in the technological efficiency and in the level of fixed investment embodied in capital facilities, rather than to adjustments in the volume of manpower or in the effectiveness of its utilization. Because the general proportioning of labor requirements to capital facilities is closely determined by technological arrangements, any gains in the effectiveness of labor efforts is more likely to reduce total labor input requirements than to increase productive capacity. Similarly, technologically induced increases in the capacity of productive facilities are likely to leave total labor requirements relatively unchanged and thus to induce a parallel increase in output per man-hour.

Processes could be characterized as "labor-dominated" to the extent that tools and other capital facilities are essentially accessories to wage earners whose manual efforts and skills represent the fundamental core of the production process—materials once again serving as the standardized inputs being acted on by manual operations: for example, custom tailoring and carpentry, various types of hand assembly and inspection operations, traditional methods of bricklaying, and many individual service occupations. And "materials-dominated" processes are those with productive capacity determined by the varying quality of the resources to which standardized labor and capital inputs are applied—encountered more commonly in agriculture, mining, forestry and fisheries than in manufacturing.

Although the differential patterning of the effects of productivity adjustments in each of these extreme cases, as well as in hybrid cases, cannot be explored in detail here,[7] it may be of interest to indicate some of the resulting hypotheses—especially because of their prospective importance in helping to differentiate the relative contributions to given productivity adjustments of different input factors.

Specifically, increases in physical output per man-hour are most likely to result from increases in the productive contributions of labor only in labor-dominated processes; in capital-dominated processes, such increases are more likely to reflect gains in the productivity of fixed capital; and in materials-dominated processes, such increases are more likely to reflect gains in the productivity of natural resources and related inputs. Similarly, increases in the productivity of fixed capital are most likely to result from increases in the productive contributions of the investment embodied in capital facilities and equipment only in the case of capital-dominated processes; in other processes, such increases are more likely to reflect gains in the productivity of the factor dominating the processes. Increases in the productivity of materials and related input are in the same way most likely to result from increases in the productivity of these inputs only in materials-dominated processes.

Viewed somewhat differently, increases in the productivity of fixed capital are more likely to increase physical output per man-hour in capital-dominated processes than in labor-dominated processes. And increases in the productivity of direct labor are more likely to increase the productivity of fixed capital in labor-dominated than in capital-dominated processes. In the same fashion, increases in the productivity of materials and related inputs are more likely to increase the productivity of direct labor and of fixed capital in materials-dominated and in extractive processes than in labor-dominated or capital-dominated fabricating processes.

Measuring the Effects of Technological Innovations Beyond Firms

Attempts to measure the economic effects of technological innovations at levels of aggregation beyond the single firm encounter new problems arising from the variety of specialized interest groups with distinctive criteria for evaluating technological innovations and their input-output effects. Within the market economy, six such groups may be identified.

1. resource suppliers (including trade unions, energy producers, raw material groups, banks, and financial institutions)
2. resource purchasers (including the procurement, personnel, and finance departments of firms)
3. resource users (including all production and fabrication departments of manufacturing firms)
4. product marketers (sales departments of companies)
5. product buyers (including firms as well as consumers)
6. government agencies (including regulators, general economic policy and special welfare units).

For example, the suppliers of productive inputs are obviously interested in ratios that relate their particular inputs to the total output of prospective users. Specifically, suppliers of coal, electricity or cotton are concerned with the effects of technological innovations on the physical amounts of these inputs required by the user per unit of his output, and reductions in such unit input requirements are usually regarded as disadvantageous by these suppliers. Providers of investment funds, sellers of machinery, and labor unions also share this emphasis on resource-oriented aspects of productivity adjustments. However, manufacturers and other users of such inputs are more deeply interested in the effects of technological innovations on the *cost* than on the quantity of such inputs per unit of output. Moreover, such users are more interested in the effect of innovations on the aggregated cost of all inputs per unit of output than on the separate costs for each input, because the users seek to find more advantageous combinations of all inputs. In evaluating his overall operation, the manufacturer is more interested in maximizing the value than the volume of his product relative to input cost—and thus is also concerned with the effects of the innovation on the quality of the product and its attendant potentials for expanding markets and fetching higher prices. And the consumer is more interested, of course, in the effect of innovations on the service values yielded by each product relative to its price than on the sheer physical volume of total output relative to input quantities and cost (Gold, 1955, pp. 13-18, 40-41, 48-54); and, more briefly (Gold, 1976, pp. 42-44).

Exploring the effects of technological innovations at levels of aggregation beyond the plant and firm suggests a need to differentiate between organically integrated and artificially contrived groupings, instead of continuing to apply essentially similar analytical frameworks to successively higher levels of aggregation—such as plants, firms, industries, and regions—without regard to the coherence of such systems. This need is highlighted partly because needed evaluations are focused on *interrelationships* among variables instead of representing single variables like output, cost, and income. And it is further emphasized by the requirement specified earlier that such ratios or comparisons can be meaningful only if they relate to an integrated system of resource flows and are derived from explicit and measurable evaluative criteria.

Possible aggregation levels cover a continuum ranging from one-man operations to the entire national economy and beyond. Among them, four basic types may be singled out for examination:

1. *Control systems:* actual networks subject to centralized control over all resource flows—as illustrated by individual firms.
2. *Specialized decisions systems:* synthetic frameworks covering those segments of actual input-output networks which bear on the decision-making evaluations of particular input suppliers (such as trade unions) or output buyers
3. *Group analysis systems:* synthetic frameworks designed to analyze the position of entire clusters of competitors with essentially similar interest—such as individual industries
4. *Descriptive groupings:* arbitrary categories lacking analytical coherence but useful as intermediate steps in computational aggregation and disaggregation—such as individual state totals for national industries.

These types are diagramed in Figure 8-5 as diversely oriented components of a generalized segment of economic activities encompassing the flow of labor,

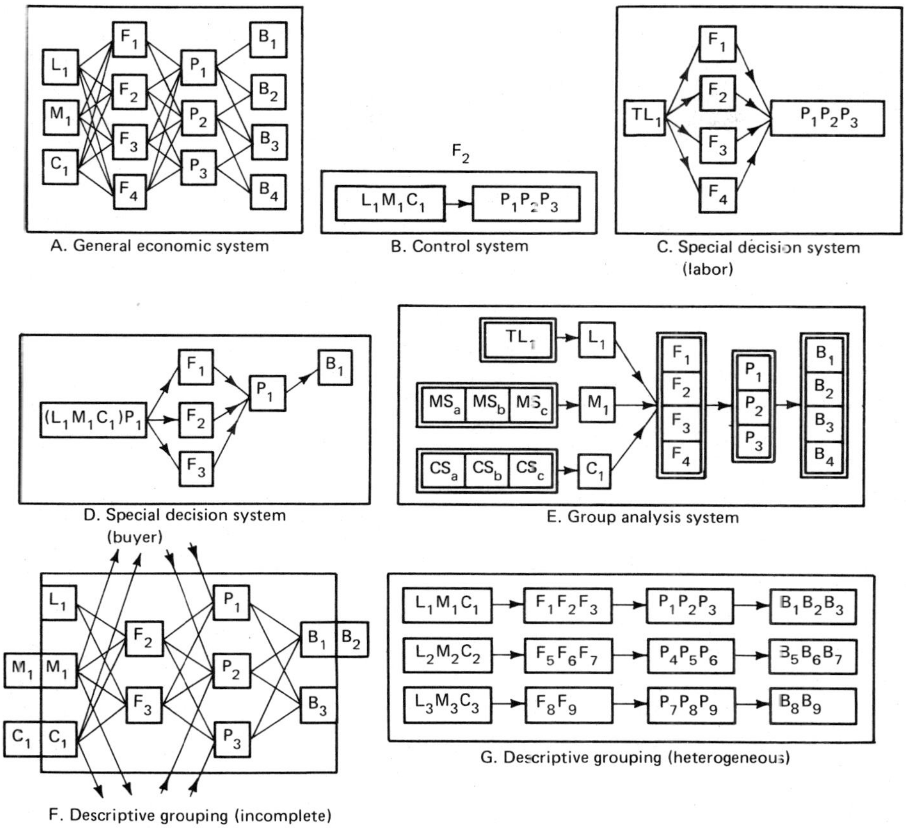

L_1L_2: Various categories of labor inputs.
M_1M_2: Various categories of material inputs.
C_1C_2: Various categories of capital inputs.
F_1F_2: Different firms.
P_2P_2: Different products.
B_1B_2: Separate buyers.

TL_1: Trade Union for L_1.
MS_aMS_b: Different suppliers of M_1.
CS_aCS_b: Different suppliers of C_1.
$(L_1M_1C_1)P_1$: All inputs for P_1.

Figure 8-5. Basic Types of Aggregated Systems

materials, and capital into several firms producing multiple products for sale to an array of buyers.

The control system is illustrated by any firm whose management controls the conversion of all inputs into outputs for maximizing specified relationships (usually financial) between total outputs and total inputs. Among *specialized decisions systems*, two kinds are illustrated: one represents evaluations by an input supplier of changes in demand for that input relative to the total output of products using it, as in the trade union model (Panel C); the other represents evaluations by a buyer of changes in total inputs per unit of output among all suppliers of the given product (Panel D). Each is analytically complete from the evaluative viewpoint of only one decision-making interest. Both *group analysis systems* and *descriptive groupings* represent multifocal aggregates, meaning that the individual firms comprising them are each subject to centralized controls and evaluative criteria, but the group as a whole is not. In *group analysis systems*, all the components have closely similar inputs and output, as illustrated by competitive firms in a homogeneous industry (Panel E). As the residual category, *descriptive groupings* represent either incomplete or heterogeneous systems. The former are characterized by leakages into or out of the production channels linking inputs and outputs, as may be illustrated by subtotals for each region of a nationally organized industry like automobile production (Panel F). Heterogeneity involves significant qualitative differences among the inputs and outputs of the firms in the group and may be illustrated by totals for each region covering all manufacturing activities, or by national totals for such heterogeneous economic sectors as agriculture or transportation (Panel G).

Multifocal aggregations pose special evaluative problems. In *descriptive groupings* which are homogeneous but incomplete, inputs and outputs may each be readily aggregated for all firms, but comparisons between them tend to be invalidated by intervening leakages. In *descriptive groupings* which are complete but heterogeneous, qualitative differences interfere even with the aggregation of the inputs or the outputs, to say nothing of allowing meaningful comparisons between them. Because they are both complete and homogeneous, *group analysis* systems such as industries permit the aggregation of inputs and outputs and also provide the congruence in their respective coverages needed to justify comparisons between them. But the interpretation of results is subject to two conditions. First, to provide an effective analytical focus, such systems must be designed in accordance with the purposes and evaluative criteria of a specified interest group: all competing firm managements, or all suppliers of a common input, or all buyers of a common product. Second, all findings must be recognized as no more than average relationships among aggregates for the group. This means that such relationships need not be applicable to actual decision-making by a particular competitor. Even more important, however, it means that such findings are difficult to interpret: one cannot automatically appraise the desirability or undesirability of a change in average input-output ratios when these may result from substitutions among input factors, changes in product mix, alterations in the relative output of different firms, or adjustments in total output. Indeed, it would be fundamentally erroneous to evaluate results in competitive industries by applying the evaluative criteria of its individual firms—such as profits—because firms often seek to enhance their achievements by undermining those of their competitors.

In short, the economic analysis of relationships between inputs and outputs may be highly vulnerable when applied to multifocal systems. This restricts the effective exploration of direct decision-making alternatives in such systems to activity sectors that can be recast into specialized decision systems. In such cases, however, each set of resultant findings would represent only one alternative interest or point of view. These alternatives could never be merged because unintegrated systems cannot have the unifying performance criteria which represent such a controlling viewpoint. And the only other level of analysis applicable to multifocal systems would be restricted to homogeneous activity sectors that can be recast into *group analysis* systems to facilitate appraisal of the general framework of pressures (or input-output relationships) affecting each group with similar interests.

Such far-reaching implications become apparent only with the realization that, beyond the level of individuals and families, control systems in Western economies are readily identifiable only at the level of separate enterprises (private firms, nonprofit institutions, and government agencies). No control systems encompass the total economy of cities, states, or regions, which usually take the form of incomplete as well as heterogeneous *descriptive groupings*. And *control systems* cover entire industries only when they have been nationalized and integrated.

On the other hand, particular regional interests might be served by synthesizing *specialized decision* or *group analysis* systems to explore input-output adjustments affecting the demand for a local mineral or for specified sectors of the local labor force, as well as others affecting the conditions of supply for certain locally consumed products. And *group analysis* systems might also be synthesized to evaluate adjustment patterns within reasonably homogeneous and competitive industries from the viewpoints of trade unions, suppliers of materials and capital, buyers of specified products, and the managers of the individual firms comprising the industry.

Some Pressures for Modifying Empirical Research Objectives

Adaptations to the Matrix of Interactions

The central objective of most empirical studies in this problem area is determining the economic effects of major technological innovations. Extensive field experience suggests, however, that this objective is not only unattainable in most cases, but that it is generally meaningless analytically. Instead, such idealized purposes often must be modified in several ways if they are to be practical targets of empirical research.

One such modification involves giving up the objective of determining the definitive economic effects of major technological innovations. Most such innovations in industry do not produce clearly defined, fixed, and measurable shifts in a set of industry-wide economic relationships. On the contrary, most such innovations trigger a continuing process of changes within an intricately integrated system of physical resource flows and associated economic valuations. It is through the ensuing complex of interactions that the innovation is "digested" by progressively more far-reaching adaptations and its effects thereby diffused through the system and over time. As a result of such a dispersion of successive impacts, and the offsetting reactions that they engender, the distinctive effects of the given innovation become increasingly indistinguishable. Hence, the objective of determining the effects of major innovations usually has to be reformulated, as has been suggested earlier, in favor of tracing the time path of the successive changes through which major technological innovations are gradually absorbed into the network of economic relationships.

A second major shift in objectives often requires giving up the effort to trace the distinctive effects of individual major innovations. Three developments tend to render such determinations impractical. First, most major innovations undergo substantial changes in form, efficiency, scale, and adaptability to specialized needs and conditions, thus posing problems concerning which stage of the evolving innovation should be the focus of such efforts. Second, most major innovations seem to be characterized by "broken edges"—that is, they seem to necessitate the development of associated trains of technological innovations as well as adjustments in products and inputs while becoming effectively utilized, thereby engendering uncertainties concerning the distinctive boundaries of the innovation whose effects are to be assessed. Third, it is difficult to disentangle the effects of a given technological innovation from the simultaneous consequences engendered by the frequent changes in operations normally introduced by managerial efforts to improve performance and to adjust to shifting market pressures. Such changes may involve significant adjustments in product design, product mix, output levels, production techniques in various sectors of operations, quality standards, and maintenance provisions. Indeed, in technologically active sectors, this often results in having several technological innovations in various stages of absorption during any extended period of analysis.

Consequently, it may frequently be necessary to shift empirical research objectives from seeking to determine the economic effects of each major technological innovation to analyzing, instead, the combined effects of the changing mix of innovations actually experienced, while continuing to make intensive efforts to identify the "core" characteristics of each innovation, to determine its direct impacts, and to estimate its diffusion rate.

Formulating realistic empirical research objectives also requires adjustments based on recognition of the difficulties of separating the effects of technological innovations from simultaneous external pressures on factor and product markets. Resulting interactions affecting the demand, supply, and prices of relevant inputs and outputs may be traced to three sources: externally induced pressures on product and factor markets that affect the attractiveness and eventual acceptance of available technological advances; the effects of such adoptions by an increasing proportion of producers on relationships in the product and factor markets; and the additional effects on such markets of competitive pressures unrelated to the innovations being studied as well as of broader economic forces.

Crude expedients have often been used, of course, to separate the effects of external market developments: most commonly involve deflating prices, either by changes in the general price level or by average market prices for particular inputs and products. But these suffer from shortcomings that are particularly critical in measuring the effects of technological innovations. In addition to shortcomings that center around the effects of innovations on the qualities and mix as well as quantities of inputs and outputs, attention must also be drawn to the all-too-frequent practice of analyzing factor and product market adjustments only after introduction of the innovations under study. But this omits consideration of already prevailing trends in such markets, which may have affected diffusion of the innovation and which would represent the appropriate levels from which differential postinnovation adjustments should be measured. For example, changes in wage rates after the innovation should be related to preinnovation trends rather than to the actual preinnovation level alone.

In short, the effects of technological innovations tend to emerge only gradually, to be overlaid by the effects of developmental improvements in the innovation itself as well as of both associated and independent technological and managerial innovations, and to interact with various external pressures on the demand, supply, and prices of relevant inputs and products. Accordingly, serious consideration must be given in formulating empirical research objectives to choosing among seeking to determine: what the effects of given innovations would have been under various unrealistic (and often irrelevant) conditions; what actual results emerged over time under whatever conditions did prevail; or some intermediate alternative of significance for clearly specified purposes.

Adaptations to Shifts in Analytical Foci

The general analytical frameworks presented in Figures 8-2, 8-3, and 8-4 are applicable at various levels of aggregation, albeit with certain modifications in coverage and the interpretation of results. At the level of the firm, Figure 8-2 may be used as presented and interpreted without special restrictions. It may also be applied without modification to individual plants that sell products to the market and have their own financial statements covering costs, revenues, and investment. Within plants, Figures 8-3 and 8-4 may be used to probe successively more deeply while retaining the capacity to reintegrate cost center results into department results and department results into plant results

The entire framework of Figure 8-2 can also be applied at the level of individual industries. But the results can no longer be interpreted as the outcome of unified managerial efforts or even as representative of the performance of individual firms. On the contrary, such measurements reflect only the average of adjustments, which may differ widely among firms characterized by dissimilarities not only in the extent to which they have adopted the innovations being studied and the effectiveness with which such innovations are being used, but also in respect to size, product mix, degrees of vertical integration, scale of production units, and so on. Such industry-level results may nevertheless be of considerable value for certain purposes relating to corporate and government policies.

Finally, the economic effects of technological innovations can be traced in a systematic way beyond the industry that adopts them by following some framework like that presented in Figure 8-6. Thus, such innovational effects on the nature, level, or mix of each of its major categories of inputs can be traced back to associated repercussions on its sources of supply and even further to the interest groups and governmental agencies concerned with the economic and social performance of that input-supplying sector. Similarly, innovational effects on the nature, design, mix, or level of the adopting industry's products can be traced forward to its impacts on the industries that buy them as well as on industries that produce substitutes for such products.

In each case, however, research efforts should begin by specifying the purposes to be served by the undertaking and using them to define the criteria to be applied and, hence, the measurements to be attempted, instead of applying any given set of criteria to all applications.

Notes

1. For further discussion, see Gold (1955, pp. 237-48) and (1971, pp. 6-12).

2. For further discussion, see Gold (1955, pp. 21-29).

3. For a more detailed discussion and specific empirical applications, see Eilon, Gold, and Soesan (1976, especially Chapters 2, 4, and 5).

4. For further discussion, see Gold (1955, pp. 18-30).

5. For more detailed discussion, see Eilon, Gold, and Soesan (1976, pp. 47-51).

6. This concept is fundamentally different from the traditional meaning of "capital-intensive." The latter is commonly applied to plants and firms with a relatively high ratio of capital costs to total costs, as in the case of electricity generating plants where such ratios may approximate 40 percent or more. But "capital-intensive" is also often mistakenly applied to operations characterized by large amounts of fixed investment and capital facilities, although these need

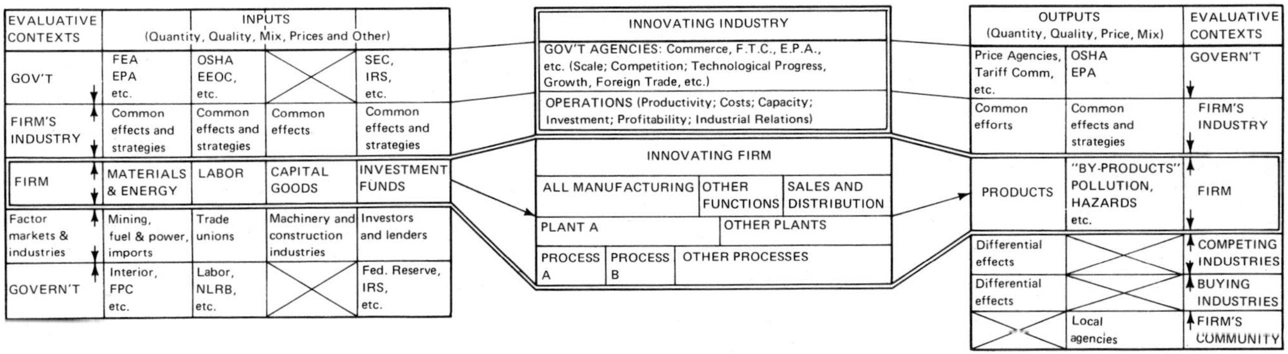

Figure 8-6. Potential Range of Contexts for Evaluating the Effects of Technological Innovations

not result in large ratios of capital costs to total costs. For example, this term is frequently applied to the steel industry because of its enormous capital requirements, although total depreciation and interest charges for major steel companies usually average 10 percent or less of total costs. In contrast, our use of the term "capital-dominated" refers instead to the extent to which productive capacity can be altered by changes in the capital facilities as compared to the effect on capacity of changes in labor or material inputs. For a more detailed discussion, see (Gold, 1955, pp. 188-196, 215-230).

7. Ibid.

References

Eilon, S., B. Gold, and J. Soesan, *Applied Productivity Analyses for Industry* (Oxford: Pergamon Press, 1976).

Gold, B., *Foundations of Productivity Analysis* (Pittsburgh: University of Pittsburgh Press, 1955).

_______, *Explorations in Managerial Economics: Productivity, Costs, Technology and Growth* (New York: Basic Books, 1971).

_______ (ed.), *Technological Change: Economics, Management and Environment* (Oxford: Pergamon Press, 1975).

Index

Index

About the Contributors

Robert N. Baird, A.B., Wittenberg University; Ph.D., University of Kentucky. Associate Professor of Economics and Senior Research Associate of the Research Program in Industrial Economics, Case Western Reserve University; co-author of *Elementary Mathematical Macroeconomics* (1971).

Myles G. Boylan, B.S., Michigan State University; Ph.D., Case Western Reserve University. Assistant Professor of Economics and Research Associate of the Research Program in Industrial Economics, Case Western Reserve University; author of *Economic Effects of Scale Increases in the Steel Industry: The Case of U.S. Blast Furnaces* (1975).

Asim Erdilek, B.A., Brandeis University; M A., Ph.D., Harvard University. Assistant Professor of Economics and Research Associate, Research Program in Industrial Economics, Case Western Reserve University; formerly Research Assistant, Arthur D. Little; Staff economist, International Monetary Fund; and Research Associate, Harvard Economic Research Project.

Gerhard Rosegger, M.B.A., University of Pennsylvania; Dr. jr., University of Graz (Austria); Professor of Economics and Senior Research Associate, Research Program in Industrial Economics, Case Western Reserve University. Contributor to *Technological Change: Economics, Management and Environment* (infra) and to A. Sichel (ed.), *Economic Effects of Multinational Corporations* (1975); co-author of forthcoming *Economic Effects of Technological Innovations in U.S. Iron and Steel Manufacturing, 1900-1960.*

About the Editor

Bela Gold received the B.S. in Mechanical Engineering from New York University and the Ph.D. from Columbia University. He is now William E. Umstattd Professor of Industrial Economics and Director of the Research Program in Industrial Economics, Case Western Reserve University. He was formerly Professor of Economics, Graduate School of Business, University of Pittsburgh; Economic Adviser in the U.S. Department of Commerce and other federal agencies; Professorial Research Fellow, Nuffield College, Oxford, England, and Visiting Professor, Imperial College of Science and Technology, London. Previous books: *Applied Productivity Analysis for Industry* (1976, co-author); *Technological Change: Economics, Management and Environment* (1975, editor and co-author); *Explorations in Managerial Economics: Productivity, Costs, Technology and Growth* (1971); *Foundations of Productivity Analysis* (1955); *Wartime Economic Planning in Agriculture*, (1949, 1969). Also co-author of forthcoming *Economic Effects of Technological Innovations in U.S. Iron and Steel Manufacturing, 1900-1960.*